T0321252

Search Engine Optimization and Marketing

Search Engine Optimization and Marketing

A Recipe for Success in Digital Marketing

Subhankar Das

CRC Press
Taylor & Francis Group
Boca Raton London New York

CRC Press is an imprint of the
Taylor & Francis Group, an **informa** business
A CHAPMAN & HALL BOOK

First edition published 2021
by CRC Press
6000 Broken Sound Parkway NW, Suite 300, Boca Raton, FL 33487-2742

and by CRC Press
2 Park Square, Milton Park, Abingdon, Oxon, OX14 4RN

Library of Congress Cataloging-in-Publication Data

Names: Das, Subhankar, 1981- author.
Title: Search engine optimization and marketing : a recipe for success in
digital marketing / Subhankar Das.
Description: First edition. | Boca Raton : CRC Press, 2021. | Includes
bibliographical references and index. | Summary: "This book explains the
nuances of Search Engine Optimization (SEO) and Search Engine Marketing
(SEM) for setting up digital websites. This book is suitable for
undergraduates, graduates and developers of digital marketing, SEO and
SEM"-- Provided by publisher.
Identifiers: LCCN 2020041982 (print) | LCCN 2020041983 (ebook) | ISBN
9780367278786 (hardback) | ISBN 9780429298509 (ebook)
Subjects: LCSH: Internet marketing. | Search engines. | Web sites.
Classification: LCC HF5415.1265 .D366 2021 (print) | LCC HF5415.1265
(ebook) | DDC 658.8/72--dc23
LC record available at https://lccn.loc.gov/2020041982
LC ebook record available at https://lccn.loc.gov/2020041983

ISBN: 978-0-367-27878-6 (hbk)
ISBN: 978-0-429-29850-9 (ebk)

Typeset in Palatino
by Deanta Global Publishing Services, Chennai, India

Dr. Subhankar Das would like to dedicate this book to his wife Prof. (Dr.) Subhra R. Mondal, daughter Ms. Gracy, and mother Mamata Das for their support.

Contents

Preface

Some psychologists say mind hunters trust, "Mind needs to perceive what it sees". Those who are resolved and have resolution to do their best to the last minute know the genuine taste of achievement. In any field, nothing can be accomplished without diligent work, regardless of whether you are running a privately owned business, independent venture, or doing work in any organization. You need to demonstrate your abilities to get the achievement. Maintaining a business isn't a simple undertaking, regardless of whether you are doing it on the web or offline. If you look at the market intently, you will locate various sites giving data about the numerous brands and their items. Web promotions demonstrate that various organizations are currently utilizing the web to connect to their intended interest group. The data for web traffic should be upgraded and sent to the specified interest group for further research and getting the best strategy out of it. You have to learn a couple of things to realize how to build up your situation in this aggressive market. You ought to likewise know to find the best solution to the question of how to end up a successful online advertiser. To get the edge over the others, you have to embrace some inventive strategies. As a matter of first importance, you need a quality site. Procuring an accomplished and adroit web specialist is of great importance, since he or she can make your site as indicated by site design improvement (SEO) that brings profitable outcomes. A decent measure of traffic to your site is fundamental and you can get the equivalent in the event that you have an SEO-based site.

Secondly, you have to monitor those destinations which are visited most by your intended interest group. When you know those sites, you'll then be able to put your promotions on their pages and the showcasing should be possible through an internet search engine. Internet promotion has some vital apparatuses, including pay per click (PPC) through which you can effectively achieve gainful outcomes. You can utilize the cost per thousand impression (CPM) to pay for the occasions your advertisement has appeared. So it is right to state that if your financial budget is huge and you continue contributing more to the information, then you can, without much of a stretch, connect with more individuals all over the world.

To interface with your intended interest group in a much better way, you ought to likewise utilize the online platform. There are numerous social locales where you can specifically advance and interface with your intended interest group. Along these lines, utilize every one of the devices of web-based promotion to help your business over the web.

Author

Dr. Subhankar Das (academician, researcher, author, writer, blogger, data science trainer, social media marketing consultant) earned his Ph.D. in Management from Siksha O Anusandhan University, Odisha, in Social Media Advertisement and Media Planning.

He is a certified professional from Google, Manipal in digital marketing, IIM Lucknow, IIM Bangalore for Media Planning, and the University of British Columbia for Advertisement, along with another 15+ globally recognized professional certifications.

He has published 30+ research papers in various national and international conferences and various international refereed journals (Scopus, Thomson Reuters, SSCI, SCIE), and even reputed IIT-IIM inhouse magazines. He acts as program committee/organizing committee/advisory committee/reviewer committee member for various international conferences across the globe. He has been a keynote speaker, session chair, and session co-chair at more than 15 international and national conferences. He is a member of more than ten international and national research societies and 10+ research organizations at a grade of senior/life. He has been awarded five international awards for teaching and research.

He has 14+ years of teaching and research experience. He is currently working as professor and researcher in the Honors Program, Duy Tan University, Da Nang, Vietnam.

He is currently working in the areas of smart tourism, AI implications in destination marketing, web 5.0, the internet of human things for digital chatbots, robotic inclusions in hospitality management, digital recruitment and acceptance model of smart artificial intelligence devices, and sustainability in the circular economy.

Introduction

Learning Objectives of the Book

Learning objectives: At the end of this book, readers shall be able to understand:

1. The concepts of SEM and SEO
2. The mechanism of how crawler and spiders work with setting up of keywords
3. Keyword generation tools
4. Page ranking mechanism and indexing
5. Concepts of title, meta, alt tags
6. Concepts of PPC/PPM/CTR
7. SEO/SEM strategies
8. Anchor text and setting up
9. Query-based search

Highlights of the Book

- **Purpose of the SEM and SEO:**

  Search engine marketing is a "form of internet marketing that involves the promotion of websites by increasing their visibility in search engine resulting pages (SERPS) through optimization and advertising". SEM includes SEO tactics, as well as several other search marketing tactics.

- **SEO and keyword relevance:**

  Your **SEO keywords** are the keywords and phrases in your web content that make it possible for people to find your site via search engines. A website that is well optimized for search engines "speaks the same language" as its potential visitor base with keywords for SEO that help connect searchers to your site. Keywords are one of the main elements of SEO.

- **Page ranking mechanism and indexing:**

  Page rank is a number to specify a position for a webpage in the search results. This number shows the exact place where a web page may come up during a particular search, how relevant it is considered to be in the given topic by the search engines. For this purpose all major search engines use special mathematical formulas – called "ranking algorithms" – to determine the rank of a webpage (Google loves to name them after various animals such as "panda", "penguin", etc.). These algorithms have an "ever-changing" feature; especially Google likes to change theirs almost from year to year. Although the exact mathematical algorithms may vary with different search engines, the principle is the same: To give a position to the examined webpage.

- **Link building:**

  Link building refers to the process of getting external pages to link to a page on your website. It is one of the many tactics used in search engine optimization (SEO). Search engines measure a website's value and relevance by analyzing the links to the site from other websites. The resulting "link popularity" is a measure of the number and quality of links to a website. It is an integral part of a website's ranking in search engines.

- **Query-based search:**

  A web search query is a query that a user enters into a web search engine to satisfy his or her information needs. Web search queries are distinctive in that they are often plain text or hypertext with optional search-directives (such as "and"/"or" with "-" to exclude).

1

Introduction to CMS, SEO, SEM, and Attribution Modeling

1.1 Content Management System

A content management system (CMS) is software that helps in content creation, editing, organizing, and publishing in different platforms. It looks for different types of editing that help in encouraging the making, altering, sorting out, and distributing of content. WordPress is a CMS that enables you to make and promote content on digitalized media [1].

Web content management is a tool that furnishes an association with an approach to oversee digitalized data on websites through the making and keeping up of content without any previous knowledge of website development or hypertext markup language (HTML). The management of web content can have helpful efficient business applications in creating bits of knowledge for decision making and delivering value-based outputs [2].

1.1.1 Phase I

In this phase, websites are developed in a simple text editor and manually edited HTML. The developer will upload files to the server as static webpages. To modify anything, the developer will edit the files and then repeat the whole process of uploading again. Since websites are having more dynamic content gradually over time, it has become prudent to have progressive user-friendly software. So, here, the main languages for web development slowly let their presence be felt for dynamic content and developers started using PHP, Perl, and similar software languages for website development.

1.1.2 Phase II

When Mambo, Joomla, Drupal, and similar frameworks entered the market, website creation accelerated drastically. You could introduce CMS on your server, select the format, and complete a website in 60 minutes.

However, and after all is said and done, we are discussing very well-informed clients utilizing these tools. You don't require a face-to-face specialist to set up a website for you; yet despite everything, you couldn't do it all alone. WordPress was normally used by either code-based website developers or those who make their own websites.

Another revolution started with the advent of high-speed Internet connections for all. More organizations needed websites all of sudden to reach out to customers online. This rapid demand brought out various differentially designed CMSs directed at amateur bloggers, specialists, and any individuals who needed to have a virtual online presence through their own websites [3].

Here, it is very important to know that WordPress, which began as a tool for blogging and maintaining small websites, is now offering a WordPress VIP package which can help to develop powerful websites. So, here, large structures and business framework CMS are slowly lagging behind in the web development market in comparison to open-source software and frameworks [4].

1.2 Types of CMS

According to Sydney Jones, Head of Marketing Communications at IXIASOFT, the CMS is a product that helps in creating, organizing, and maintaining digitalized content. Generally, CMSs bolster multiple clients and give adequate advantages to large companies by cost saving and increasing cooperation between technically expertized teams. This is very helpful in taking total control of the content [5].

So, an understanding of different CMSs is the key factor in selecting the best available alternatives for the business. There are five mainstream CMS explained in Figure 1.1.

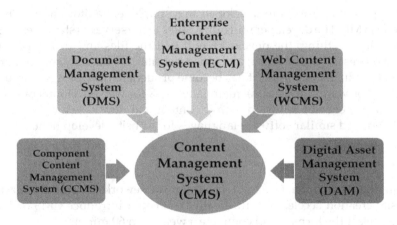

FIGURE 1.1
Different types of CMS available.

1.2.1 Component Content Management System (CCMS)

A component content management system (CCMS) varies from a standard CMS on the basis of identifying and differentiating content at a fundamental level. Rather than managing content page by page, it takes words, expressions, passages, or photographs (otherwise called "segments") and stores them in an archival repository. Intended for maximum content reuse, parts are just stored once. The CCMS functions as a reliable, confided-in content management that distributes content over different platforms like mobile view, PDF, and print.

1.2.1.1 Advantages of CCMS

- **Reusability:** Content reuse inside a CCMS saves time during the composition, altering, and publishing stages, and essentially diminishes interpretation or translation costs.
- **Traceability:** A CCMS empowers you to track content in detail. You can see who did what, when, and where.
- **Single-sourcing:** With a CCMS, you can push content to numerous channels, including print, mobile view, web view, chatbots, etc.
- **Enhanced team collaboration:** Improve the work process for your content development team working remotely.

1.2.2 Document Management System (DMS)

Paper is practically wiped out. Following business records on paper is a relic of times gone by. So here DMS offers a paperless answer for managing, storing, and tracking archives on cloud platform. It gives a digitalized answer for transferring, managing, and sharing business data without any issues of duplication, printing, or scanning.

1.2.2.1 Advantages of DMS

- **Eco-friendly:** Digitalizing management of business data saves a lot of paper, which is highly eco-friendly.
- **Security:** DMS provides differential levels of incremental privacy to all the data and content.
- **Mobile-friendly:** With this DMS, one can access the documents on mobile and even can edit them.

1.2.3 Enterprise Content Management System (ECM)

ECMs gather, arrange, and convey an enterprise's documentation, guaranteeing that the basic data are conveyed to the right stakeholders of the organization (workers, clients, business partners, etc.). It also provides access to content to all individuals of the enterprise. It also erases records after a specific maintenance period, ensuring that no pointless content will take up space on website.

1.2.3.1 Advantages of ECM

- **Flexible:** It helps in getting any file in any form, and processing and storing it conveniently.
- **Increases efficiency:** As documentation storage and management are handled well, it makes the system robust and efficient.
- **Decrease in cost of storage:** It helps in storing the important documents, erasing the unnecessary data.

1.2.4 Web Content Management System (WCMS)

This helps the client in managing the digitalized segments of the website without having any prior knowledge of coding or programming. It manages the content by providing help in coherent collaboration, writing, and managing tools. It only handles the web content, unlike other styles of content management which manage both web and print.

1.2.4.1 Advantages of WCMS

- **Personalization:** It provides customized design and content to modify website pages.
- **Automaton:** It saves time and helps in qualitative productivity with automated content publication.
- **Scalable:** It enables organizations to develop exponentially without thinking about the over-usage of website data limits.

1.2.5 Digital Asset Management System (DAM)

This helps clients to store, arrange, and share developed content easily. It offers a basic centralized library where all stakeholders can get developed content. These advantages incorporate sound, innovative records, video, archives, and introductions. A DAM is cloud-based, so clients can access content from anywhere.

1.2.5.1 Advantages of DAM

- **Centralized repository:** Content is protected and secure in one repository.

- **Effective brand management:** A DAM enables you to deal with a marked online interface for clients to get significant files.
- **Digital publishing:** With a DAM, you can push digitalized content to outside managers who have no direct link along with online channels and portals.

1.3 Website Marketing

Website advertising is the method of marketing the contents and promoting so that more traffic can come to the page, consequently promoting the product, increasing the perceivability among the visitors and brand advancement, and lastly resulting in a SALE [6].

1.3.1 Need for Website Marketing

1. **It's advantageous:** Website promotion of a product helps in empowering open dynamic business without any time constraints in the opening and closing of physical establishments. Consumers can reach out to the website at their convenience.

2. **Mass reach:** With mass appeal and reach to diverse consumer segments, and promotion on a website, the marketer can overcome all physical obstacles like place, area, and geographical segments. The marketer can their reach to every nook and cranny of the market.

3. **Cost:** Promotion is very cost-effective as compared to physical promotion in on-site physical stores. It reduces cost by minimizing stock keeping and physical display of items in store. Just-in-time and kaizen principles help in minimizing the overhead cost.

4. **Personalized selling/strategically pitching:** Internet promotions help in giving customized service to the clients by knowing their purchase intentions and behavior. Marketers are better placed to deal with consumers' history and inclinations. Judging the traffic, the website marketer can focus on the offers that ignite the purchase intention. Websites can better analyze the data collected from interested consumers' clicks and how they look for their convenience along with patterns of purchase. These give a better insight for all the actions of website advertising.

5. **Trends and knowledge:** As huge traffic comes to the website, there would be a tremendous measure of information streaming into the framework [7]. Breaking this information down for patterns and experiences would be simple. Example: When to change from selling summer stock to winter products as per the market.

1.3.2 The Eventual Fate of Web Promoting

1. Google's AMP Initiative.
2. Latent semantic indexing keywords.
3. CRM applications.
4. Interactive content and voice search.
5. Mobile marketing.
6. Schema mark-ups.
7. Influencer marketing.

1.4 Digital Branding

Internet branding or digitalized marketing is an approach to get more pro-motion for your product image on all platforms of web-based propagation, particularly on search engines and social media. The website developer takes web-based branding to the next level by making each online existence of the brand a definitive image. It goes past pretty much making a blog or an Internet-based account on various social media, since, overall, in excess of 500,000,000 individuals like or share at least one brand via web-based net-working media.

1. Website browsing along with use of the Internet is developing in leaps and bounds such that today it has become an integral part of our daily chores.
2. More than half of web-dependent consumers follow at least one brand on the web as they want to:
   a. Be updated with all recent news.
   b. Avail of the opportunity to win a lucky draw or to have some emotive benefits.
   c. Be unable to avoid uncommon offers.

Online client–brand communications have become progressively prevalent among (potential) clients. For organizations, one objective group is most sig-nificant: Brand fans! After clients move toward becoming fantastic apprecia-tors of the brands, they do the following actions:

- Frequent visits to the brand shops.
- Spreading positive traits of the brand by electronic word of mouth (e-WOM).

- Spending a lot of money on brands as they are brand loyal, as opposed to clients who did not progress toward becoming brand fans.

Via web-based networking, social media fans of the brand end up being excited, faithful, and brand-situated. In any case, these tremendous quantities of online posts are not really countable or controllable for organizations despite the fact that they do impact your image and related notoriety! The main concern is by what means can organizations measure or potentially facilitate essential client–brand cooperation (to the extent that this is conceivable and attractive)?

1.4.1 MOST Framework to Address Web Advertisement

The framework which is very commonly used for web advertisement that solves the marketer's needs is called **MOST**:

M: What do clients associate **MOST** with brands via web-based networking and social media?

O: To **OBJECTIVELY** quantify valuable data from online client–brand cooperation.

S: What is the **STRATEGY** to create an impact on client–brand associations for firms?

T: How can website administrators use **TOOLS** on Internet-based life brand pages?

Client–brand connections via web-based networking media emerge from clients' and organizations' thought processes:

a. Customer–brand connections that emerge without an organization's help, for instance online audits, web journals, and articles.

b. Customer–brand connections that emerge from firm activities, for example, the Internet-based life brand page. Clients associate with the firm by tapping on "like" or posting a remark.

1.4.2 Insights into Perceived Brand Image and Brand Loyalty for Web-Based Advertising

1. The analytics of web crawlers for very big quantities of online item surveys give insight into engaging brand experiences for creating a favorable brand image.

2. Web analytics helps in analyzing the brand image according to brand affiliations regarding the changes with positioning and repositioning of the brand.

3. Managers can impact client communications with their image pages via web-based networking media by:
     i. Asking questions.
    ii. Organizing challenges.
   iii. Adding recordings.
    iv. Placing them at the highest point of your page.

1.4.3 Digital Media

This is commonly known as online media that collaborates and includes photos, videos, and music, which either have intellectual property rights protected or not protected [8].

1.4.4 Types of Digital Advertising

Digital advertising has come a long way in terms of appearance, resonance, and reach from the initial beginning of a static pop-up image to dynamic moving ones. Currently there are seven types of advertising to be found in all digital platforms.

1.4.4.1 Display Advertising

The first type of web-based promotion, these are visual advertisements that show up on outsider sites normally identified with either substance or administration. These advertisements have developed from the essential type of pennant promotions. These days, these advertisements come as:

a. **Images:** These are static fundamental flag advertisements. They show up in and around the product and features.
b. **Content of text:** Content promotions only promote the in-depth advertisements significant to the product's content.
c. **Float promotions:** These move over the screen or float over the website's content about the product.
d. **Promotion by wallpaper:** These are the display advertisements and change the basic view of a website and cover the complete page.
e. **Promotion by pop-up:** These are altogether new windows that pop-up before the opening of the main website. They show complete promotion for guests who browse the webpage.
f. **Promotion by flash:** These are dynamic diverse ads which last for minimum time and have maximum impact on giving an idea to the consumer.
g. **Promotion by video:** These are automatic auto-played tools of promotion where the content is in video format. Marketers assume that, due to trust and loyalty, the consumer will prudently play later.

These advertisements have a moderate effect on the outsider website and its legitimacy. Websites charge for the traffic and have a minimum basic rate for the advertisement. Outsider destinations, similar to the Google Display Network, take all those relevant social media focused advertisements into consideration, all of which help you focus on the crowd that would be well on the way to be keen on your product.

1.4.4.2 Social Media Advertisements

Online social networking ads and their promotion preview aren't just proficient, however powerful. Fundamentally, advertisements on display network and social media platforms may be either an image or an automatic playable video [9]. Web-based life publicizing is incredible in light of the fact that you can focus on your group of spectators splendidly. For instance, Facebook is focusing on age, district, interests, and basic idea. This phenomenon is like an atom of the whole plethora of web advertising platforms. Social media advertisements (SMAs) are of two types:

a. **Organic SMAs:** They build trust and give input from the interested group, like e-WOM.
b. **Paid SMAs:** They influence the developed content of the product by payment and contact explicit individuals.

The best platforms to target are:

a. Facebook for promotion and top of channel showcasing. *Please follow Facebook's Ad Guidelines and find the algorithm for predicting the profits of advertisements and campaigns.*
b. LinkedIn for B2B deals.
c. Twitter.

Different platforms to hit up if clients have limits of expenditure are Google+, Pinterest, Instagram, Tumblr, and Reddit. So, one can set up one's own social media adverts by oneself to show the organization's content. In the next section, the symbiotic aspect of SEO is described, known as search engine marketing (SEM).

1.5 Search Engine Marketing (SEM)

This is the most trustworthy, reliable, widely used digital paid form of promotion. It works with help of catchphrases – which keywords of one's organization and similar ones attract traffic with the main objective of getting the website higher on the search engine results page (SERP). These promotions show up in different search engines and their content [10]. These are of two types, pay per click (PPC) and cost per thousand (CPM).

a. **PPC:** These offer keywords and their outcomes which show the highest point for SERP in search engines. It is a bundle of benefits to the client, as with payment you can set the individual keywords on the adverts. This is relatively easy to set up.

b. **CPM:** These are also known as click per impressions (CPI). The client will be charged at a level rate of 1,000 impressions. It is based on a budget and its spending limits for the websites which ensured impressions based on clicks clicks on SERP.. If there are no clicks, then it becomes a problem for SERP set-up. One cannot evaluate the performance until the setup is finished.

One can use SEM in an unpaid way by managing the keywords, which is known as search engine optimization (SEO). Here the web crawlers keep an eye on unpaid results for improving the SEO and subsequently get more clicks. The best platforms of SEM, like Google's AdWords, help to focus more on ranking and SERP. There are 13 ways to increase your Google Ads click-through rate (CTR) that don't cost a cent and sweep the ultimate guide to Google Tag Manager to benefit as much as possible from AdWords ventures. SEM for Bing is less challenging than AdWords.

1.6 Native Promotions or Advertising on Social Media

Localized promotion is supported by sponsorship for various Facebook pages and presented on other digital platforms. They are incorporated and covered to the stage where they can set up [11]. This can be done by native advertising through TOI, ET, SAKAAL, etc. Various types of native promotions are found in general, like in-feed advertising, search advertising, recommendation gadgets, and affiliated adverts.

1.7 Remarketing/Retargeting

This is the method with the most potential for marketing toward individuals who for some reason are loyal to and affiliated with products and management of all sorts. Retargeting is very commonly used for popular products too. It relies on those who are conversing with the website. When consumers visit the website, if some eye-catching trivial offer is given to them as a sweepstake like free air tickets or free gifts, then they will be pretty happy about it. And while browsing your website for gifts or some attractive features/information, the advertisement will run repeatedly for the items and products. This is very cost-effective and follows the principles of attention, interest, desire, and action (AIDA). If these types of promotion are done properly, then they are more rewarding than PPC. It helps in building changes in attention, creating interest, making the items more desirable for consumers, and obviously bringing the purchase intention into action. Facebook and Google remarketing are only a few methods, apart from these, there are lots of things which the marketer can use as an outside way to set up the remarketing activities.

1.8 Advertisement by Videos

Apart from YouTube which is the most potent, prevalent, and surely understood of all video adverts, there are a lot of other video platforms where promotions can gather terrific momentum. You can go the course of instructive/educational content. Or, on the other hand, perhaps you need to present a how-to. A content developer can try to bring an interesting and attractive story by creating innovative adverts. It is perfect for building superlative branding, particularly in the event that you have a product that is mostly shown to browsers for generating interest. Video ads are picking up in prevalence since they maintain a strategic distance from unmitigated promotion while additionally pulling in the restricted capacity to focus of numerous creative content developers who create and post the content to YouTube, Facebook, Twitter, Vimeo, Bright roll, YuMe, Hulu, AOL, etc. Short pre-roll adverts are also helpful in creating awareness which run before the actual video content commences.

1.9 Marketing by Electronic-Mail (E-Mail Marketing)

Since the advent of Internet promotion, e-mail promotion is the least expensive, quickest, and most dynamic method of promotion one can have. It is the most sophisticated client attraction tool and helps in promoting the deals by effective use of e-mail campaign managers such as MailChimp, ConvertKit, Campaign Monitor, Active Campaign, AWeber, Constant Contact, and Get Response. This way, the message of promotion can efficiently spread and have a good return on investment (ROI). For the setup of a campaign, first a summary of the e-mail is prepared by various tests or by attaching a newsletter on the website. Then, the e-mail is sent on headways, cutoff points, features, or content introduced in the blog. Messages need to be short, simple, and precise so that they can reach and hit the augmentation pattern of consumer thinking. So, both SEO and SEM act in conjugation to have a synergistic effect on the web display of the advertisement.

1.10 Search Engine Optimization Concept

It is a a new method for content and its promotion where the Internet is concerned. It started with the production of web crawlers themselves. The primary known development of a web crawler was back in July of 1945 by a Dr. Vannevar Bush. He utilized the idea of hypertext and memory expansion to accumulate a group of researchers to set out and cooperate to help fabricate an assortment of information for all humankind. He was not exclusively a firm devotee to putting away information, yet he additionally accepted that if the information source was to be helpful to the human personality, we ought to have it speak to how the customers or visitors view visitors mind attempts as well as could be expected or perceived [12,13,14]. He at that point proposed the possibility of, for all intents and purposes, a boundless, quick, solid, extensible, cooperative memory stockpiling and recovery framework. He named this gadget a MEMEX. A couple of decades later, Gerard Salton and his groups at Harvard and Cornell built up a program called the Salton's Magic Automatic Retriever of Text (SMART) educational recovery framework. Gerard also mentioned this in his book called *A Theory of Indexing* which clarifies a significant number of his tests which search engines can be seen to be still generally dependent on.

This in the long run prompted the absolute first web crawler which was named "Archie". The program was made by Alan Emtage of McGill University in Montreal. The initial couple of sites started in around 1993; yet the majority of these used Archie. The web crawler assisted with taking care of configuring a link for all scattered information by linking a content-based information gatherer that had an ordinary articulation matcher for recovering document names coordinating a client question or query. Google didn't come on the scene until September of 1998 and was initially made by Larry Page and Sergey Brin while they were PhD students at Stanford University, in California. It wasn't long after this that web indexes began turning into typical easily recognized names as more families started to engage with the Internet. This was after two students (Page and Brin) at Stanford distributed a paper called "The Anatomy of a Large-Scale Hypertextual Web-Search

Engine". The archive contained data with respect to "PageRanking" which is a framework that web-search tools today, for example, Google, still use to help rank list items dependent on quality, and not watchwords or key identifiable words alone.

The mid-2000s were the time, when Google gradually started its takeover and to set the rules of fundamental SEO. The way to deal with the web being something beyond words began to truly come to fruition in November 2003 as the "Florida" update to Google's calculation occurred. Websites had started to lose their rankings and Search Engine Watch considered the reaction to Florida a huge "objection". It is essential to take note that numerous sites profited by this occurrence too online. It was the principal episode in which websites were punished for similar keyword stuffing, showing Google's suggestions for the client first – for the most part with quality substance. 2005 was a major year for the improvement of SEO. This was the year that Google joined with Yahoo and MSN for the Nofollow Attribute. This was to help decrease the measure of nasty connections and remarks on sites. In June of that year, Google introduced customized search. Customized search utilized the user's inquiry history to help make their outcomes increasingly pertinent [15]. Today, individuals are as yet growing better approaches to help improve the utilization of SEO to ensure that the individuals who show up at the highest point of query items are by and large genuinely positioned. Ensuring that your organization's site is upgraded to its fullest with significant watchwords and content will enable you to climb the exceedingly significant SEO ladder and help you get higher traffic moving through to your site (Figure 1.2).

SEO TIMELINE

1945	1965	1993	1998	2003-05
Dr. Vannevar Bush utilized the idea of hypertext and a memory expansion to accumulate solid, extensible, cooperative memory stockpiling and recovery framework called MEMEX.	Gerard Salton and his groups at Harvard and Cornell built up a program called the SMART (Salton's Magic Automatic Retriever of Text) educational recovery framework.	First web crawler which was named 'Archie' made by Alan Emtage of McGill University in Montreal.	Larry Page and Sergey Brin archived contained data with respect to 'Page Ranking' which is a framework that web search tools use today.	SEO is used with Page Ranking by Google, MSN etc.

FIGURE 1.2
SEO timeline.

1.11 Method of PageRank

This is the method which is used by Google for deciding the significance and importance for a website page. This helps in indexing of the page. It is widely known as *Google Juice*. It was created by Larry Page and Sergey Brin at Stanford during their research in 1998. The name was coined by Larry as his surname is supposed to be used here. Before that, all webpages were indexed by keyword importance and one could play with them by using similar words and rehashing them from time to time so that the rank would go up. Developers sometimes put hidden content on pages to rehash too.

1.11.1 Benefit of PageRank Measurement

a. *PageRank attempts to measure a webpage's importance.*

Google follows the assumption that the most significant pages have the maximum number of connecting links. This technique uses links as votes and connects the selected page. References are interlinked with the points and the significance is discovered. The frequency of iteration is mostly what is significant to all. This bodes well since individuals will in general connect to pertinent content and pages with more links for accessing the complete progressive creative information. PageRank never stops at link ubiquity and takes into account additional content for signifying linking pages. A webpage with a higher rank will have more weightage in voting with their connecting links. It also additionally helps in getting connections on pages which can be eligible for voting. Pages with more irrelevant connections have less weightage too. Significant, important pages get better traffic and become sources which is not the case with less important pages.

b. *Importance of PageRank.*

It is one of the best tools for ranking the webpage in terms of generating clicks and generating links. It significantly affects the Google rankings for a webpage.

1.11.2 Demerits of PageRank

The prominent demerit is that a high-rank developer can control the information. An exception to PageRank control is the use of Google bombs. So, for this, Google is taking logical steps to make the positioning true. "Connection cultivating" also controls the PageRank. Connection cultivating is the act of connecting without pertinence. It is the method of connecting those pages which are normally mechanized. Arbitrary connections help in giving a connection pool or links. Google adjusts the figures of traffic to channel off conceivable connection pools. So, it is always advisable not to present or register websites for low ranking links. Try not to freeze when it is connected to the connection ranch, since time will have either no or less impact on position, so that developer cannot control the connections in any way.

1.11.3 Visualization of PageRank

This is calculated on a 1–10 scale and a number is allotted to each individual page inside the website. Not all pages will get 10 out of 10 on quantity. Not many pages have a PageRank of 10, particularly as the number of pages increases.

1.11.4 Need for Improving PageRank

Many individuals accept that PageRank doesn't make a difference regarding traffic, and that is valid. There is no immediate connection between PageRank (PR) and traffic; however, having a decent Google PageRank increases the believability of a site. Website admins with involvement in online business understand the advantages of a PR. There are numerous publicists hoping to support a blog entry on a blog; however, they pay special attention to high PR. On the off-chance that your PageRank is high, you can charge more for advertisement spots and support ship posts. Besides, when different bloggers see your webpage with a great PageRank, they won't fret about connecting to you next time, as they will consider you to be a trustworthy source [16].

Google PageRank is a value that distinguishes each page of the webpage dependent on the significance of the Internet page. You can improve your PageRank by utilizing the accompanying thoughts:

a. *By using unique and search engine optimized content.*

   The most significant way is to make exceptional substance for your website or blog. Web-search tools love great quality and specific content which is refreshed all the time, and the web indexes will crawl your webpage all the time. The crisp precise content of your webpage will draw in different website admins to your website and that will give you valuable one-way inbound connections. Likewise, after Google's ongoing post calculation, it is essential to have a blog with a better than average refreshed post each month otherwise the website positioning can be dropped. Keep up a post recurrence and ensure you spread your substance by means of online networking channels [17].

b. *Using heading tags and keywords.*

   Having quality precise noteworthy content is significant, and yet you should be found by visitors on Google and the most ideal method for getting that going is by putting yourself on page 1. *This should be possible by – Keyword inquire about > SEO upgraded post > Social media advancement > Backlinks.* The most urgent part is improving your substance for specific keywords.

c. *Backlinks.*

The most significant factor in the activity of improving the PageRank of a site or a blog is the inbound connections to the website. Truth be told, the accomplishment of the incomparable Google PageRank is exceptionally subject to the quantity of inbound connects to a page. Not all connections are equivalent with regards to backlinks. Connections from high-PR websites are more significant than low-PR sites, yet these inbound connections should originate from important websites to beneficially affect the PR [18].

d. *Article directory submission.*

Compose top-notch articles for the well-known and high-PR article catalogs with a connection back to your site in the end of the article at the resource asset box. In some cases, these articles will be taken by individuals for distribution on their site and that will give you a valuable one-way inbound connect to your site. Google thinks about these connections as well-known decisions in favor of your blog or webpage, and thusly they will enable you to expand your Google PageRank [19].

e. *Submitting blog to web directories.*

A straightforward method to improve the PageRank of the website is to present the webpage or the blog to a high web index. In the event that you present your site to more registries, you will get more backlinks, which will expand your site's odds of positioning highly.

f. *Asking other bloggers to backlink.*

There is nothing incorrect in telling different bloggers about a greatly admired or read article you composed. For instance, in the event that you have composed an extraordinary investigated article about getting a higher PageRank with zero backlinks, you can recommend it to different bloggers who expound on SEO and you may very well get some great backlinks [20].

g. *Fixing broken links.*

This is something you can do to expand the Internet searcher's amicability of your blog. You can utilize modules like Broken Link Checker for WordPress, or broken connection checker sites to recognize every messed-up connection and fix them. You can likewise utilize Webmaster instruments to locate every single broken connection and pages connecting to it, and work on changing the connections or setting up 301 redirections. There are numerous approaches to build page positioning on Google, and the significance of having a high PageRank ought not be belittled. A high PageRank can enable you to expand the validity of your site/blog and help you increment your income as well [21].

1.12 Google Search Engine Results and Actions

Every web index uses a three-way method for solving, supervising, positioning, and returning issues. None has a hint of the background activities to generate clicks and queries. So, search engines work out how to present the content and their applicability for general queries. Three important functions are needed for this. They are as follows:

a. *Web crawling.*

This is used to find out the content keywords given on the Internet. Normally duplication and replication are seen. So, the uniqueness of keywords is what can give maximum benefit [22]. Example: Robots, crawlers, creepy crawlies, or web crawlers.

b. *Indexing.*

When a spider or web crawler enters a webpage for indexing, duplication is not helpful and the server goes with the unique keywords. So, the crawler makes the duplication for websites. Google positions them far and wide in all aspects and technical structures. The name file acts like a storage for information. Files are structured in a particular order and arranged in priority of information for the search inquiry question [23].

c. *Algorithm.*

When web crawlers have come to one place on the website, then they need to be refreshed so that search will not be hampered. For this reason, they need to be positioned properly. Here the *algorithm* is important. The mathematical calculation is what makes the system effective as a searching and information-providing tool. Web crawlers and web index use large algorithms to be always refreshed and get top position. These lead *SEOs to encourage site owners and SEO elements to climb in the rankings.* Web indexes answer the queries put forward by a search merely depending on keywords. They give results in form of links where SEO helps in generating the SERP [24].

1.12.1 Relevance and Popularity of Search Engines

The important goal of finding a page is the responsibility of a web crawler. Initially they were pretty much oversimplified and poor in execution. But now developers have contrived better approaches to deal with search keywords. Now, web indexes are accepting a page or webpage or archive more prominently [25]. There is always a doubt about client satisfaction with keywords, but the significance is not achieved. But the monitors use calculations to sort out what to keep and what not to put in content so that ubiquity is not compromised. So, this is the way that web crawlers rank pages, and this is called *Ranking of Page (PageRank)*.

1.13 Attribution Modeling

Attribution is a way of distinguishing a lot of visitors' activities ("occasions") moved over screens and contact focuses that contribute in some way to an ideal result, and after that doling out an incentive to every one of these occasions or events [26]. The attribution model is defined as a set of rules or standards which decide on various ways of allocating benefits to touchpoints or clicks caused by transformation due to deals done by developers [27]. For instance, the Last Interaction model in analytics doles out 100% credit to the last touchpoints (i.e., clicks) that quickly go before deals or transformations [28].

1.13.1 First-Touch Attribution

As this model gives benefit on the basis of single touch point, it acts to overemphasize a single channel. So, it helps in giving more importance to promotion, which then gives more meaning to advertisement. It is very difficult to attain, but this can be vulnerable due to its favor toward innovation confinements [29] (Figure 1.3).

1.13.2 Lead-Generation Touch Attribution

This model overtakes the first-touch model sometimes when this attribution model is unknown/absent or the guest didn't round out a lead structure. The advantage of this model is that it helps in showing the channels which can bring changes. It is significant, but on the other hand it's only a little piece of the total client venture. In a B2B structure, advertisements are targeted for 100% lead generation with this attribution type [30] (Figure 1.4).

1.13.3 Last-Touch (Opportunity Creation Touch) Attribution

Here in this model, there is very little time to allocate the promotional attributes to various business opportunities. So, the research platforms have the shorter first touch and the lead creation is also very small. This attribution

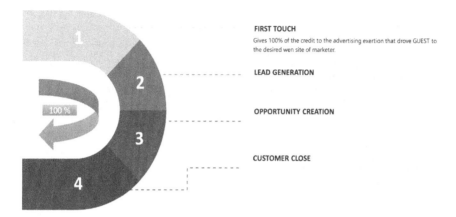

FIGURE 1.3
First-touch attribution.

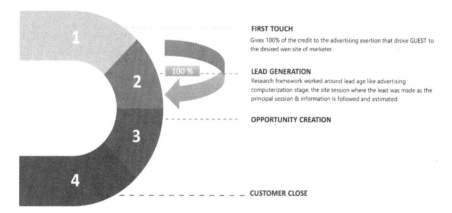

FIGURE 1.4
Lead-generation attribution.

does a 30-day termination process for advertisement. On the day of assigning benefits, the last click is given all the credit of 100% [31] (Figure 1.5).

1.13.4 Last Non-Direct Touch Attribution

Website research and direct information are very much in demand. Traffic ascribed directly to the website is regularly under the eyes of promotion

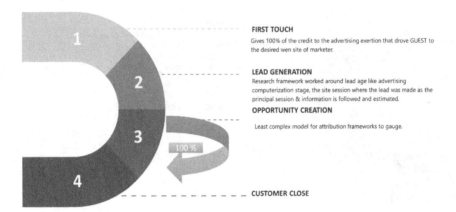

FIGURE 1.5
Last-touch attribution.

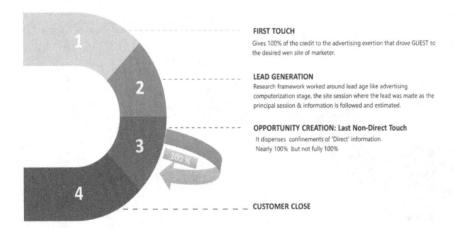

FIGURE 1.6
Last non-direct touch attribution.

researchers and developers for new clicks. Non-direct touch signifies that the click comes indirectly through the various SMAs. Here direct information is deceptive. The significant upside of the last non-direct touch, at that point, is that you keep away from the inconveniences of direct channel information [32] (Figure 1.6).

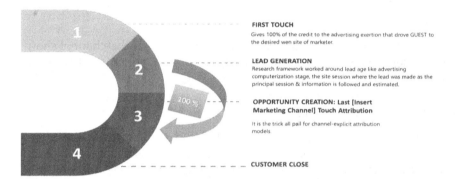

FIGURE 1.7
Last insert marketing channel touch attribution.

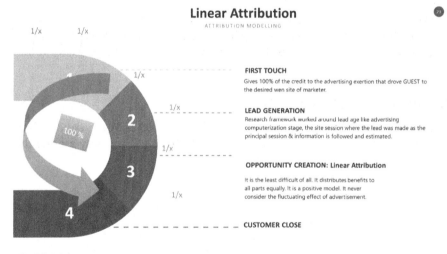

FIGURE 1.8
Linear attribution.

1.13.5 Last (Insert Marketing Channel) Touch Attribution

Developers will need to utilize the last AdWords Touch Model for showing their worth. They will utilize Facebook or Twitter's last-touch models. Here, all models are one-sided for their own channel, and may sometimes exaggerate their own individual effect [33] (Figure 1.7).

1.13.6 Linear Attribution

Figure 1.8 shows a linear conversion attribution model, where conversion values are assigned to the campaigns, mediums or keywords that help in generating conversions in proportion. Google analytics by default doesn't use the linear attribution model.

1.13.7 Time Decay Attribution

This will never give a decent measure of credit to top-of-the-channel show-casing endeavors since that will consistently be the most distant from the change [34] (Figure 1.9).

1.13.8 Position Attribution

This shows groups that give attention to the lead age that track each touch-point and gives more importance to the initial one at first click and the third one at lead creation for 40% credit each. In the middle stage till the lead is generated it gives 20% credit. It is U-shaped [35] (Figure 1.10).

1.13.9 Position Extended Attribution

This model takes the extension of position from the first click to opportunity creation. It is W-shaped, every stage has an equal share of 30% credit at the beginning and 10% of profit is distributed in each stage [36]. It is very much helpful in all aspects until closure (Figure 1.11).

1.13.10 Full-Path (Z-Shaped) Attribution

In this model, each of the touchpoints at the four key stages gets 22.5% of the credit and the last 10% of profitability is distributed similarly among the remaining touchpoints (Figure 1.12).

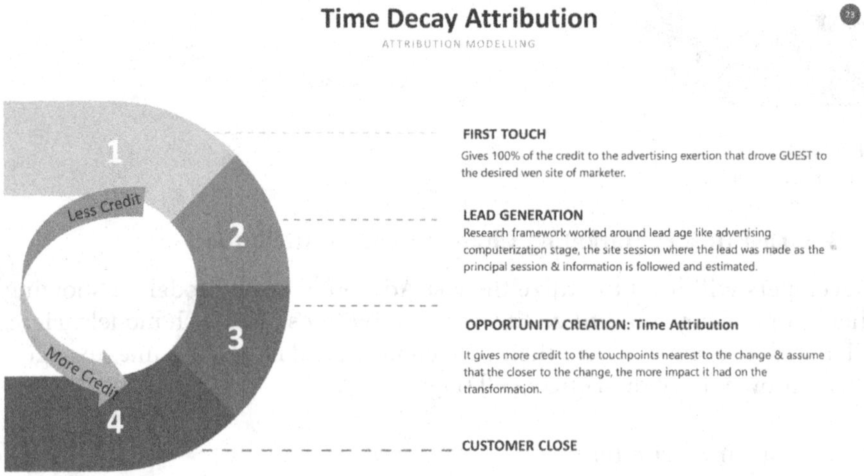

Time Decay Attribution

ATTRIBUTION MODELLING

FIRST TOUCH
Gives 100% of the credit to the advertising exertion that drove GUEST to the desired wen site of marketer.

LEAD GENERATION
Research framework worked around lead age like advertising computerization stage, the site session where the lead was made as the principal session & information is followed and estimated.

OPPORTUNITY CREATION: Time Attribution
It gives more credit to the touchpoints nearest to the change & assume that the closer to the change, the more impact it had on the transformation.

CUSTOMER CLOSE

FIGURE 1.9
Time decay attribution.

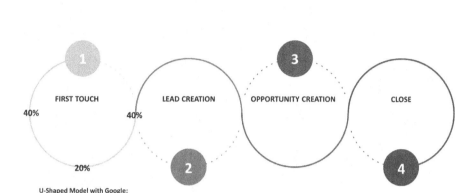

FIGURE 1.10
Position attribution.

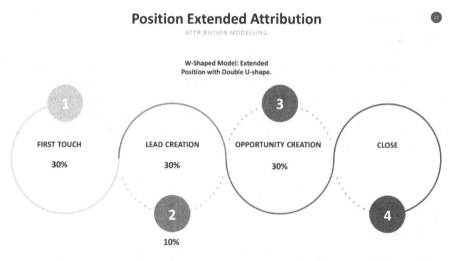

FIGURE 1.11
Position extended attribution.

1.13.11 Customized Algorithmic Attribution

This adjusts to the client's requirements to the developer's procedure. It breaks down the client's information to show the outsized effect of specific advances. It is the most difficult and tedious model of all to assemble and use. But it has the power to give the desired output of precise input from the client.

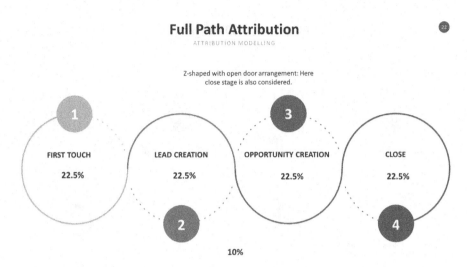

FIGURE 1.12
Full-path attribution.

1.14 Conclusion

This section shows how SEM, alongside SEO, functions with PageRank technique and how it puts the CMS into training; it discusses the need for attribution, demonstrating in content for website and social media marketing executives. In the contemporary world, web indexes are the rudder of all substance and the algorithmic setup is given in the ensuing parts, for example, site essentials and on-page, off-page setups.

References

1. Cahill, K., & Chalut, R. (2009). Optimal results: What libraries need to know about Google and search engine optimization. *The Reference Librarian, 50*(3), 234–247.
2. García, R., Verdú, E., Regueras, L.M., De Castro, J.P., & Verdú, M.J. (2013). A neural network based intelligent system for tile prefetching in web map services. *Expert Systems with Applications, 40*(10), 4096–4105.
3. Hou, D., Chen, J., & Wu, H. (2016). Discovering land cover web map services from the deep web with javascript invocation rules. *ISPRS International Journal of Geo-Information, 5*(7), 105.
4. Huang, C.Y., & Chang, H. (2016). GeoWeb crawler: An extensible and scalable web crawling framework for discovering geospatial web resources. *ISPRS International Journal of Geo-Information, 5*(8), 136.

5. Katumba, S., & Coetzee, S. (2017). Employing search engine optimization (SEO) techniques for improving the discovery of geospatial resources on the web. *ISPRS International Journal of Geo-Information, 6*(9), 284.

6. Lopez-Pellicer, F.J., Florczyk, A.J., Nogueras-Iso, J., Muro-Medrano, P.R., & Zarazaga-Soria, F.J. (2010). Exposing CSW catalogues as linked data. In Marco Painho, Maribel Yasmina Santos, and Hardy Pundt (eds.), *Geospatial thinking* (pp. 183–200). Berlin, Heidelberg: Springer.

7. McGee, M. (2010). By the numbers: Twitter vs. Facebook vs. Google buzz. *Search Engine Land*, available at: http://searchengineland. com/by-the-numbers-twitte r-vs-facebook-vs-googlebuzz-36709 (accessed on 17 November 2019).

8. Onaifo, D., & Rasmussen, D. (2013). Increasing libraries' content findability on the web with search engine optimization. *Library Hi Tech, 31*(1), 87–108.

9. Singh, S., Mondal, S., Singh, L.B., Sahoo, K.K., & Das, S. (2020). An empirical evidence study of consumer perception and socioeconomic profiles for digital stores in Vietnam. *Sustainability, 12*(5), 1716.

10. Singh, L.B., Mondal, S.R., & Das, S. (2020). Human resource practices & their observed significance for Indian SMEs. *Revista ESPACIOS, 41*(07). Retrieved from http://www.revistaespacios.com/a20v41n07/20410715.html

11. Sharma, E., & Das, S. (2020). Measuring impact of Indian ports on environment and effectiveness of remedial measures towards environmental pollution. *International Journal of Environment and Waste Management, 25*(3), 356–380. doi: 10.1504/IJEWM.2019.10021787

12. Das, S. (2020). Innovations in digital banking service brand equity and millennial consumerism. In Kamaljeet Sandhu (ed.), *Digital transformation and innovative services for business and learning* (pp. 62–79). Pennsylvania, PA: IGI Global.

13. Mondal, S.R. (2020). A systematic study for digital innovation in management education: An integrated approach towards problem-based learning in Vietnam. In Kamaljeet Sandhu (ed.), *Digital innovations for customer engagement, management, and organizational improvement* (pp. 104–120). Pennsylvania, PA: IGI Global.

14. Das, S., Nayyar, A., & Singh, I. (2019). An assessment of forerunners for customer loyalty in the selected financial sector by SEM approach toward their effect on business. *Data Technologies and Applications, 53*(4), 546–561.

15. Subhankar, D., & Anand, N. (2019, May). Digital sustainability in social media innovation: A microscopic analysis of Instagram advertising & its demographic reflection for buying activity with R. In *1st International scientific conference "modern management trends and the digital economy: From regional development to global economic growth" (MTDE 2019)*. Atlantis Press.

16. Tricahyadinata, I., & Za, S.Z. (2017). An analysis on the use of Google AdWords to increase e-commerce sales. *SZ Za and I. Tricahyadinata (2017) International Journal of Social Sciences and Management, 4*, 60–67.

17. Singh, I., Nayyar, A., Le, D.H., & Das, S. (2019). A conceptual analysis of internet banking users' perceptions. An Indian perceptive. *Revista ESPACIOS, 40*(14), 1–17.

18. Mohanty, P.C., Dash, M., Dash, M., & Das, S. (2019). A study on factors influencing training effectiveness. *Revista Espacios, 40*, 7–15. Retrieved from http:// www.revistaespacios.com/a19v40n02/19400207.html

19. Singh, I., Nayyar, A., & Das, S. (2019). A study of antecedents of customer loyalty in banking & insurance sector and their impact on business performance. *Revista ESPACIOS, 40*(06), 11–28.

20. Gupta, D.K., Jena, D., Samantaray, A.K., & Das, S. (2019). HRD climate in selected public sector banks in India. *Revista ESPACIOS, 40*(11), 14–20.
21. Singh, S., & Das, S. (2018). Impact of post-merger and acquisition activities on the financial performance of banks: a study of Indian private sector and public sector banks. *Revista Espacios Magazine, 39*(26), 25.
22. Jain, S., Jain, V., & Das, S. (2018). Relationship analysis between emotional intelligence and service quality with special evidences from Indian banking sector. *Revista ESPACIOS, 39*(33), 3–16.
23. Das, S., Mondal, S.R., Sahoo, K.K., Nayyar, A., & Musunuru, K. (2018). Study on impact of socioeconomic make up of Facebook users on purchasing behavior. *Revista Espacios, 39,* 28–42. Retrieved from http://www.revistaespacios.com/a18v39n33/18393328.html
24. Mondal, S., Das, S., Musunuru, K., & Dash, M. (2017). Study on the factors affecting customer purchase activity in retail stores by confirmatory factor analysis. *Revista Espacios, 38,* 30–55. Retrieved from http://www.revistaespacios.com/a17v38n61/17386130.html
25. Mondal, S., Mall, M., Mishra, U.S., & Sahoo, K. (2017). Investigating the factors affecting customer purchase activity in retail stores. *Revista ESPACIOS, 38*(57), 22–44.
26. KumarSahoo, K., & Mondal, S. (2016). An analysis of impact of electronic customer relationship management (e-CRM) on service quality of e-retail stores: A study of Bhubaneswar. *INDEXED BY, 10,* 10–12.
27. Mondal, S., & Sahoo, K.K. (2020). A study of green building prospects on sustainable management decision making. In Arun Solanki and Anand Nayyar (eds.), *Green building management and smart automation* (pp. 220–234). Pennsylvania, PA: IGI Global. doi: 10.4018/978-1-5225-9754-4.ch011
28. Das, S., & Nayyar, A. (2020). Effect of consumer green behavior perspective on green unwavering across various retail configurations. In Vannie Naidoo and Rahul Verma (eds.), *Green marketing as a positive driver toward business sustainability* (pp. 96–124). Pennsylvania, PA: IGI Global. doi: 10.4018/978-1-5225-9558-8.ch005
29. Nadanyiova, M., & Das, S. (2020). Millennials as a target segment of socially responsible communication within the business strategy. *Littera Scripta, 13*(1), 119–134. doi: 10.36708/Littera_Scripta2020/1/8
30. Patil Swati, P., Pawar, B.V., & Patil Ajay, S. (2013). Search engine optimization: A study. *Research Journal of Computer and Information Technology Sciences, 1*(1), 10–13.
31. Pellicer, F.J. L., Béjar, R., & Zarazaga-Soria, F.J. (2012). *Providing semantic links to the invisible geospatial web* (Vol. 1). Zaragoza: Universidad de Zaragoza.
32. Purcell, K. (2011). Search and email still top the list of most popular online activities. *Pew Internet & American Life Project, 9,* 1–15.
33. Vockner, B., & Mittlböck, M. (2014). Geo-enrichment and semantic enhancement of metadata sets to augment discovery in geoportals. *ISPRS International Journal of Geo-Information, 3*(1), 345–367.
34. Mustafa, R.U., Nawaz, M.S., & Lali, M.I. (2015). Search engine optimization techniques to get high score in SERP's using recommended guidelines. *Science International, 27*(6), 5079–5086.
35. Nebert, D. (2004). Developing spatial data infrastructures: The SDI cookbook v. 2.0. *Global Spatial Data Infrastructure, 2,* 39–56.
36. Ochoa, E.D. (2012). *Analysis of the application of selected search engine optimization (SEO) techniques and their effectiveness on Google's search ranking algorithm* (Doctoral dissertation, California State University, Northridge).

2

Website Basics and Development

2.1 Introduction to Static and Dynamic Websites

2.1.1 Static Websites

These contain constant fixed content where every page is coded in HTML and shows data to the visitor. They are basic fundamental websites with no web programming. These static websites work with the help of HTML pages and web server [1]. Smaller websites with little or constant content can run smoothly with fixed code. Static content will not be suitable for dynamic content and larger websites. So, static content makes the job of developer easier to refresh and format [2] (Figure 2.1).

2.1.2 Dynamic Websites

Dynamic websites are costly at initiation but mostly preferred due to the scope for many opportunities in the future. Fundamentally they allow the developer to refresh and add content at own convenience, for example, events associated with the organization can be shown as a direct program interface. Often creativity blocks the dynamism of content. A few significant uses are content management, e-commerce business, information sharing, organizations' intranet and extranet, online record keeping and transfer from one portal to another, and dynamic customization as per higher authority to the existing content [3]. They operate with data that frequently changes in nature depending on the viewer's time zone, time of browsing, language, and what type of information is sought after. It may go with dynamic content client or server-side scripts and sometimes with both, also depending on HTML programming for the basic structure. Dynamic content determines the strength and how the traffic can come more to the website (Figure 2.2 and Figure 2.3).

Numerous recent websites from the last ten years are mostly static, however an ever-increasing number of individuals are understanding the upside of dynamic websites and what kind of scope they can offer. Dynamic websites have double-edged benefits, for example, either the developers can take

STATIC WEBSITE
Advantages & Disadvantages

PROS

Rushes to create

Cost Effective

Simple

Good For Small Websites

Simpler for web indexes to record

Quicker to move on moderate associations

CONS

Costly more for long run.

Hold up until developer makes schedule-wise to roll out the improvements.

Limited usefulness

Preparation may be sometimes tedious, costly and progressing.

FIGURE 2.1
Static website advantages and disadvantages.

DYNAMIC WEBSITE
Advantages & Disadvantages

PROS

Functionality is higher

Easy to Update

New update content pulls traffic & helps in SEO

Act as a collaborative system for users & staff

CONS

Takes time in developing

Expensive & slow to react at introduction

Hosting cost is more

FIGURE 2.2
Dynamic website advantages and disadvantages.

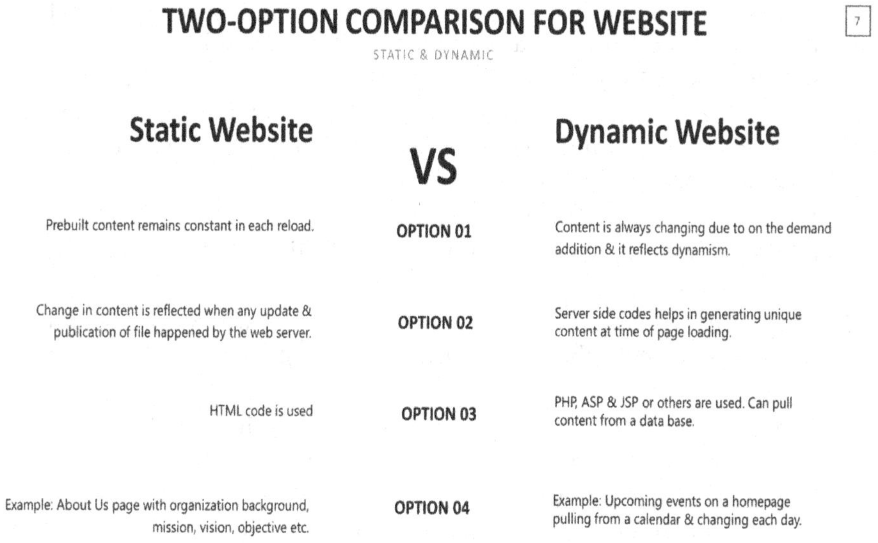

TWO-OPTION COMPARISON FOR WEBSITE

STATIC & DYNAMIC

Static Website	VS OPTION	Dynamic Website
Prebuilt content remains constant in each reload.	OPTION 01	Content is always changing due to on the demand addition & it reflects dynamism.
Change in content is reflected when any update & publication of file happened by the web server.	OPTION 02	Server side codes helps in generating unique content at time of page loading.
HTML code is used	OPTION 03	PHP, ASP & JSP or others are used. Can pull content from a data base.
Example: About Us page with organization background, mission, vision, objective etc.	OPTION 04	Example: Upcoming events on a homepage pulling from a calendar & changing each day.

FIGURE 2.3
Comparison between static and dynamic websites.

advantage of websites or they can create an enriching experience by employing the content as a tool for setting up an intriguing background for visitors' objectives [4].

2.2 Platforms Used to Build Websites

2.2.1 WordPress and CMS Employability

WordPress represents an open-source CMS from October 2009 that helps developers in developing dynamic websites. Mullenweg and Mike Little first designed and published WordPress on 27 May 2003. CMS generally stores each and every piece of information like pictures, content, and reports for accessing on the website [5]. Also, it allows modifications, adjustment, and distribution of content on the website.

2.2.2 Features

1. Client management: The process of client management allows the developer to use the client's data and change the jobs assigned from website developer or writer to edit, add, or delete client password or data. Clients can only authenticate the content.

2. Media orientation: The process of media orientation deals with records of media and organizes them for customized transfer, sorting, and maintaining them for future use.

3. Background theme management: WordPress helps in viewing the new edit, background flipping, where images, templates, records, and pages are used for a better view.

4. Plugins extension: Client customization is only possible with the addition of useful plugins and extending their capacities as per requirements.

5. SEO management: SEO operation also helps in a few site designing improvements that enable the search engine to optimize and become direct.

6. Language orientation: Language set-up helps in translating the whole content into the language preferred by the client.

7. Information support: WordPress also provides support by customization, documentation, remarking, post paging, and adding labels.

2.2.3 Pros and Cons

The pros and cons of using WordPress are shown in Table 2.1.

2.2.3.1 Hypertext Markup Language (HTML)

HTML is the language for developing webpages and helps in connecting HTML reports. So, the connection accessible on a site is commonly called hypertext. Developers use HTML for increasing web content with labels since it helps in web programming. HTML was used for supporting structure, headings, records, etc. It also helps in data sharing with logic and their interpretation. Nowadays, HTML helps in organizing pages with labels or

TABLE 2.1

Pros and Cons of WordPress Use

Pros	Cons
1. Open-source free platform, CSS documents can be customized.	1. The use of several modules sometimes makes the site heavy and inaccessible.
2. Numerous modules can be accessed by developers and clients for easy customization.	2. PHP information is often required for making any changes to a website developed by WordPress.
3. Extremely simple using a What You SEE Is What You Get (WYSIWYG) user interface.	3. Sometimes refreshing is bit tedious and may result in loss of information.
4. Effective rapid media transfer is seen, helping in SEO direct establishment and helping in completing various jobs for clients.	4. Modification and alteration of media files and tables can be troublesome sometimes.

tags. HTML is a *markup* language and utilizes different labels to arrange the substance. These labels are encased inside content and its accessories. The greater part of the labels have their corresponding closing labels. *For instance, <html> has its end tag </html> and <body> tag has its end tag </body>, and so forth* [5].

2.3 Hypertext Pre-Processor (PHP)

Previously this was known as personal home page (PHP) and was open-source as clients found it very useful and suitable. In 1994, Rasmus Lerdorf introduced the first use of PHP. Then slowly it evolved to a pre-processor to hypertext and acts as a server-side script for language that is used in HTML. It is used to develop dynamic content and databases from entire web-based activity for business destinations. It coordinates with MySQL, Oracle, and Microsoft SQL server for its operation. It is actively executed by Apache module on Unix. MySQL helps in getting solutions to complex queries with great outcomes in record time. It strengthens real-time conventions like POP3, IMAP, LDAP, PHP4, etc., by Java and transmitted objects designs like COM and COBRA. It also helps in improving probability by n-level of improvement. PHP also acts as a tool for pardoning as it is used as method of escaping the negative links. The PHP syntax looks like C language. PHP has five characteristics: It is simple, efficient, secure, flexible, and familiar with the setup around the world. It is used in the following ways:

1. It helps in structure where PHP collects information from different sources and gives different types of data.
2. It helps in the removing of, and alteration in content which leads to a greater outcome for client.
3. It treats the factors and subfactors as per their access.
4. It also helps in controlling the traffic on the website.
5. It can encode the source information.

2.4 Domains, Hosting, and Related Concepts

Web developing helps in enriching the content. Website facilitation is an individual option since it incurs a lot of costs if the site is developed without a site like Yahoo or Google. Commonly, the developer buys server space from an internet service provider (ISP) to host [6].

2.4.1 Developers' Hosting Platforms

Developers normally use two types of hosting platforms:

1. **Windows**: If one has affection for Windows, then, at that point, the developer finds facilitating servers maintaining various kinds of hosting platforms from which space can be purchased. These are expensive as they give programming along with server engagement.
2. **Linus**: This provides unlimited opportunities to the developer. It is less expensive than Windows. ISPs prefer to provide these hosting platforms like *Unix*.

2.4.2 Types of Hosting

For a developer, there are seven types of hosting available, from which, depending on budget and plan, the most effective one is selected [7] (Figure 2.4).

1. **Free Hosting**

   Some organizations allow their web servers to host other websites for free so that they can run their commercials on the websites [8]. They offer free space in exchange for allowing their advertisements to run. The website developer doesn't have to pay, they only need to approve. Examples: *lycos.com, myspace.com*, etc., who give space to include website pages.

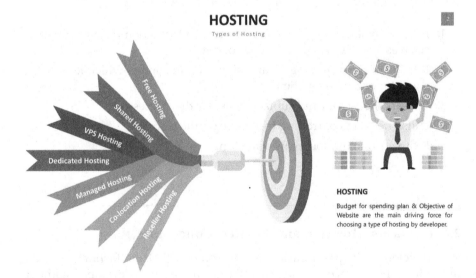

FIGURE 2.4
Types of hosting.

2. **Shared Hosting**

   Here, the website is facilitated by a high-speed server along with other different sites. The client will have its own login and password for mutual hosting and is permitted to work in customized territory. There is no option for connecting any document or registry [9]. Indeed, you would not even know how many destinations are facilitated on your common host. This kind of hosting facilitation is very practical and useful for smaller sites where space and speed are insignificant. Traffic on one side affects the speed of other collaborating websites.

3. **Virtual Private Server Hosting**

   This is a sort of virtual private network (VPN). It is one of most widely used for website hosting where virtual techniques are used for customized dedicated client service on a server for many users [10]. It is basically a sever running inside a server. A physical server hosts many isolated individual virtual servers where hypervisor software makes divisions and keeps everything separated. Each virtual host can be rebooted independently.

4. **Dedicated Hosting**

   This is basically the same as a virtual private server (VPS). But here the total server will be assigned to the website. They are expensive and should be monitored at times of heavy traffic [11].

5. **Managed Hosting**

   This is an information technology (IT) provisioning model where the server provider leases space to the host. It gives hardware for hosting to developers [12]. Basically, it works for the developer and client in its own way.

6. **Co-Location Hosting**

   All servers cannot give everything to a website all the time. So, the requisites are often mutually symbiotically hosted in other locations [13]. Co-location hosting acts as an alternative way for availing developers' cooperation in all facilities. If high traffic is expected then this hosting will be highly beneficial.

7. **Reseller Hosting**

   Here, the web hosting provider permits a few or all the web services bundled in a pack to be sold by a third-party seller. It helps the organization to act as a server without being able to control the architect of web hosting.

2.4.3 Elements of Hosting

There are elements of hosting on which value creation will be dependent when the developer purchases a space for a website from a server providing the facility. This value creation is organization-dependent and collaborates with the different segments associated with it.

1. **Space on Disc**

   An average medium-sized website will require somewhere in the range of 10 to 100 MB of space on disc, in the event that you intend to upload a ton of audio and video. A prerequisite of buying server space is to grow an alternative disc space which might be required for contingencies.

2. **Traffic for 30 Days**

   An average medium-sized website will require somewhere in the range of 1 GB to 10 GB of information spread on a month-to-month basis. In the event that you intend to keep a great deal of sound and video on your site, at that point you need an arrangement with a higher information storage limit. Check various choices dependent on your requirements. What are different alternatives accessible in the event that the developer goes beyond the permissible space so that the website does not crash?

3. **Processing Speed**

   In the event that you are purchasing space on a common machine, at that point you can't think about how much speed will be given to you. All things considered, this path is to see other facilitated destinations with a similar specialist co-op to think about their facilitating quality. In any case, if a developer is purchasing an online server or dedicated server, at that point you must think about the amount of RAM speed required. So, the purchase decision will depend on the handling capacity for the website.

4. **Connectivity Speed**

   Connectivity speed is the most sought-after issue for a website. So, the association speed for connectivity should ideally range between 64 kbps and 2.48 Gbps.

5. **E-Mail Accounts**

   A sufficient amount of e-mail accounts on platforms like IMAP and POP along with forwarding facility is very much helpful for a website. The client will need e-mail to communicate with potential customers via the server-customized domain ID which gives the client a unique position and customization.

6. **E-mail Support**

   Support for the e-mail interexchange from the website server to the potential visitors by the client is very much needed for smooth maintenance of the correspondence. E-mailing support is mostly back end support but potentially the most frequently demanded work for a website. Simple mail transfer protocol (SMTP) servers are needed for smooth setting and correspondence of web servers [14].

7. **New Technology**

   New technology always comes with support for website maintenance and can deal with PHP, ASP, Java, and so on. This gives and
   sets a trend of credibility and compatibility with new tools and
   image in customers' eyes.

8. **Databases**

   Databases like MySQL, Oracle, and SQL Servers help the server-
   dependent website to work properly. If space is purchased on mutual
   servers then website will depend on the amount of content that the
   website needs. ISPs generally don't provide excess space. So devoted
   space for the database controls this prerequisite.

9. **Uptime for Server**

   All developers need 99.9% uptime from the server providers. If the
   server is down then there should be alternative options for navigation in a website. Uptime is the direct opposite of downtime. So,
   server uptime has to be at its best.

10. **Reinforcement and Backup**

    Backup and reinforcement play a very important role in smooth
    maintenance of the website. Backup is what is needed for ordinary
    websites as it will help in restoration if a problem like server crash
    happens; otherwise it is of no use as such. It helps in the overall man-
    agement and operation of the website.

11. **Panel for Control**

    One has to check for a better control panel, or similar instrument,
    that the developer is providing, which will help in better effective
    facilitation. It helps to have the option for keeping up essential activ-
    ities identified with your site, for example, logging your administra-
    tion demand, your reboot demand, or some other issue.

12. **Customer Support**

    Before settling an arrangement with your specialist organization,
    you should ensure they give you the needed help. Developer can use
    the data on online forums for discussions. Specialized organizations
    exist that also provide round-the-clock customer support which can
    be outsourced.

2.5 Do's and Don'ts of Website Making

When you are making a site (or enlisting a web/blog developer to make
one for you), there are explicit things you should know about. Generally,
for personal websites and professional blogging, specific significative things

are needed. At first developers may act very much on their own but later in the process they always act with respect to the client's requirements so that maximum traffic will come [15]. So here is a list of dos and don'ts for a developer for the effective designing of a website.

Do: *Make structured webpages.*

Of late there is a structure demand for webpages found in the market based on *CSS* documentation and lattice design [16]. Pages should be lighter, and content shouldn't be cluttered; data should be spread out and followed. *Example: 960.gs.*

Don't: *Never include boxes.*

Some new developers put 20+ boxes of random sizes and single paragraph for all data in the content at one place. This should not be done as it lacks arrangement. This is termed a hurricane of madness. If a developer can't arrange all content in a systematic fluid manner, then they can't be called a designer or developer.

Do: *Area of focus and importance of identification.*

It is safe to say that you are developing a website for a product-centric business. Assuming this is the case, ensure that the landing page is focused on the product. Then in the suggestions, the product should be the important objective [17]. If it is a personal blog which is focused on a complementary gift or likewise to pull traffic, then ensure the blogs are getting the best possible place for measurement and consideration. Sites like WOO Themes work admirably and advance what their principle center is – WordPress themes.

Don't: *Put unnecessary ads on the page.*

In case you're going to attempt to profit from your site/blog, help yourself out and lay off the over-the-top commercials. If your webpages are 70% ads and 30% content, then chances are high for visitors to leave and likely not return. So, an optimized mix of ads and content is what the pages need.

Do: *Select the right color for webpages.*

A color plan is essential for effective visualization and how the webpage will be seen by the visitors. You won't need a brilliant and "uproarious" shading plan for reflective webpages. Mostly websites go with a CMYK plan for pink or yellow mix, while the more specialist websites mostly go with lighter color options.

Don't: *Clutter with 20 different colors.*

If a developer uses a 64 set of all colors then it looks awful and very much drives the visitors away from the website. Contents will be mixed up and importance will go down. So, a site like *Color Lovers* can help developers in optimizing the webpage color pattern and selecting a plan for five color combinations, so that the webpages will be properly visualized.

Do: *Easy scan of webpage.*

Visitors take a maximum of five minutes to identify whether they need to explore more or not on a website. So, the correct data and content at the right place and the right frame will drive visitors into more exploration. It is recommended to use H labels (horizontal way of representation) for getting maximum content highlight. It can be achieved easily by using pictures, tables, images, and quotes.

Don't: *Write lengthy paragraphs.*

If a developer puts up a post of more than 1000+ words compiled in a single paragraph, then the whole blog will not attract visitors. So, content needs to be precise and fluid. So bigger content needs to be simplified and made smooth.

Do: *Make it simple to sign up.*

Visitors like few options in the sign-up structure. If a developer has cluttered the framework with multiple options, then it will likely have less sign up. Visitors prefer a quick and simple structure with an easy framework.

Don't: *Describe endlessly.*

Drifting, unreasonable content with a large number of smiley emojis and irregular descriptions will make it difficult for visitors to focus. You need them to remain – act like it. So, one can't understand properly due to lack of important information or with use of smiley emojis in the blog or webpage. It will drive away visitors [18].

Do: *Get a proper copywrite of quoted content.*

Where the developer knows that words matter the most, it will be prudent to keep them short, precise, and simple. If some words for key essential representation are borrowed from previously used content, then the developer should get a copywrite of previously published content. So, choosing the right phrase, keyword, and page headings helps in improving the content up to 90% to attract visitors [19].

Don't: *Populate content with keywords.*

Neither visitors nor Google are fools. If keywords are used in large numbers without having any significance, then it will lead to bad traffic or fewer visits. So, the composing of content is paramount [20].

Do: *Make navigation easy.*

Navigation is what makes the website easy to access. So, a basic route can be indicated by a blue color and sign up can be given by green. But in every sense, the developer has to make the route navigation very simple to spot and use.

Don't: Make the search process tedious.

Visitors should be given 30–40 seconds for getting their desired information from the content. If they have to navigate for any longer, it is unlikely that they will return again. So, the use of spotlight information is what is always recommended.

Do: Optimize your load times.

In the event that there's one repeating subject in this whole chapter, it is that guests are fretful. You have to assemble your site with the ideal browsing speed and help the pages to load in 1–2 seconds and enable your page to stack in around 1–2 seconds with the help of *CSS* documents, JavaScript, and Google so that they will be structured for the ideal websites.

Don't: Put all the content as images on the webpage

If the developer puts all the content in image form, then the authenticity of content will go down and it can't be viewed properly. If the content has an image of more than 1 megabyte (MB) of memory, then it is very difficult for the browsers to see. Also, if diverse JavaScript is used then also the content will be burdened by 10+ modules of content. It will take more than 20 seconds to download important information [21].

Do: Select appropriate fonts and font sizes.

Typography is an exceptionally significant part of website architecture. Use of correct segment titles and text styles will significantly impact the experience your guests have when surveying your sites [22]. As a rule, you should utilize one primary text style for the substance, and after that the developer may change titles to alternate textual style.

Don't: Use different fonts and font sizes.

It is highly recommended not to use five different fonts and ten different sizes in the content. It looks clumsy and not eye catching. Uniformity and fluidity of content will be missed terribly.

Do: Give visual appeal to the webpages.

Visual appeal is what helps attract the visitors to an image of the website. Website content composition attracts maximum views and clicks. Introduction and surface inclination maintain the flow. So, visual attraction is the most important aspect of website design and development.

Don't: Throw a bunch of vague content together and think you'll do well.

Animation-centric gifs are your first no-no. They make the content cluttered and scrambled that take after a multi-year-old Myspace page [23]. It isn't charming, and if you're not mindful of it, then the website may not get many visitors. Things have changed, and individuals don't hope to see something that resembles old-style content representation. In case you're an expert, keep this at back of your mind and ensure your webpage structures are adequate [24].

2.6 Tips for Making a Great Website

In the online age that we live in, having an excellent website is an outright need for just about all promotion decisions and business adventures. An extraordinary site can all the while work as a promoting device, a store stage, a presentation of works and abilities, a correspondence channel, and as a motor for branding in the virtual world that creates interest in visitors' minds.

2.6.1 Strategize Your Brand

The primary undertaking you have is to come to a nitty gritty understanding of the product branding procedure. You would prefer not to begin making a site before you have a strong idea of the site's motivation, its target group, its visual content that attracts browsers, or its manner of representing the data [25]. These will to turn out to be obvious to you as you seek after the following:

1. **Identify your target market**: Who are they? Which statistic gatherings do they have a place with? What do they like to do? How would they see themselves? In what manner will your business or site impact their lives?

2. **Research your competitors**: A developer should research what the competitors are posing as a challenge to the website, what their qualities and shortcomings are, and how you can find a specialty for yourself in the field.

3. **Define your brand identity**: On the off chance that you needed to portray your image in three words, what might they be? What is the image that comes to mind? Think about your image's character and concentrate from its substantial qualities like hues, jargon, and style [26].

4. **Prepare consistent branding material**: Presently this is a great opportunity to get pragmatic and set up the materials that will before long be highlighted on your site and on different stages as well. For example, a logo, pictures, trademarks, recordings, printed substance, and the sky is the limit from there. They should all relate to your brand image and serve your brand positioning.

2.6.2 Approach the Design

1. **Choosing a template**: Website formats give you a strong establishment for planning. Preferably, the format you pick as of now has the design you need and is styled by your conceived branding methodology.

2. **Customizing your template**: When your heart is set on a format, it's an ideal opportunity to transform it into your own one of a kind site by tweaking and altering: Including your very own customized content (pictures, writings, backlinks, recordings, sound) and refining the structure if necessary (images, shapes, page request, textual style decision, and the sky is the limit from there). Your representation can be as fundamental or as detailed as you need it to be [27].

3. **Tweaking the look**: There is considerably more to making a website than simply picking pictures. You can mix it up with highlights that will upgrade your site configuration, similar to parallax, looking to add dimensionality to your site that infuses it with dynamic movement, wide content coverage wide strips to emphasize the page format considerably.

2.6.3 Prioritize Usability

Your site, as dazzling as it seems to be, should likewise give a good experience to your site visitors. A lovely site that does not work appropriately won't get you far. As you make your site, focus on these significant points:

1. **Navigation flow**: Ensure the website and its pages are clear and instinctive with the goal that guests can undoubtedly explore among pages and subpages, utilizing the principle menu or inward connections.

2. **Content hierarchy**: Intelligent content and a chain of command direct the visitors through your site in the way that best serves to your advantage. The most essential perspectives ought to be the most unmistakable, and the plan must plainly mirror that. The landing page is a really genuine case of content pecking order done right [28].

3. **Call to action (CTA)**: CTAs are the messages that invite site guests to make a direct move, similar to: "Register Free" or "Get Yours Today". So, they tell the guests unequivocally what it is that you need them to do. Figure out how to make the ideal CTAs before you add them to your site.

4. **Readability**: What's the purpose of having a site if your content is essentially disjointed? Ensure you utilize clear textual styles and agreeable text dimensions, that your content can balance well with the basic structures, and that you have enough "void area" around your writings.

5. **Footer**: The base piece of your site is known as the "footer" (the top is the "header"). Footers are not immediately noticeable to visitors; however they can be utilized in various approaches to improve ease of use, for instance: Include the majority of your content data there, including links connecting to online live channels; show a streamlined website map that connects to all pages of the webpage; compose a brief About Us section, or webpage disclaimer content.

2.6.4 Prepare for Search Engines

Getting your site to rank unmistakably on query items is one of the most significant approaches to building your traffic, which is the reason it's so imperative to combine search engine optimization (SEO) as of now when you make a customized website. Inside web-based advertising, SEO is a science in its very own right, and its key components are:

1. **Keyword research**: Place yourself in the shoes of your potential site visitor or customer. What will they search on Google that should lead them to your site? The catch words that they will utilize are the ones that will control your SEO technique. Settle on an educated choice on which of these keywords you must target.

2. **Text**: Your Bio area, your blog, your footer, your FAQ segment – each identifies that has content ought to be conceptualized in light of SEO. The real adventure for a developer is to locate an unpretentious and rich approach to coordinating your catch words with your site's content without compromising on quality. Web search tool crawlers are shrewd, and in the event that they think you sound too much like an advertisement, they will down-position you [29].

3. **Meta tags**: Meta information isn't obviously noticeable to your site guests. Web crawlers, in any case, do peruse your meta information, and you can control what it is that they see and how they present your webpage in query items. Pursue this manual to see how to utilize meta labels to further your potential benefit.

4. **Alt text**: Web crawlers can't peruse pictures (yet!), yet that doesn't imply that pictures are pointless for SEO. Each picture that you transfer to your site ought to have an "Alt Text" added to it. This *short line* discloses to web crawlers what the picture portrays, which thus enables your pictures to be "found" in query items.

5. **Link building**: If all links are represented properly, your indexed list positioning improves and you have different sites connecting to your site. You can begin by presenting your site to indexes, ensuring that the majority of your web-based profile information point to your webpage, and urge website guests to share your content too.

Remember that SEO is a long-haul process that does not end once your site is moved up. You should keep on refining it as you go so as to win manageable outcomes slowly.

2.7 Responsive Website

Compatibility and responsiveness are two dimensions on which the success of a website depends heavily. Compatibility with different operating systems and platforms is highly needed from developers. A website should operate in operating systems like Android, Windows, and macOS without any glitch. Webpages should be responsive to different platforms on various devices like iPhone, iPad, Android phones, Kindle, and other phablets [30]. New innovations demand new changes in responsiveness. The primary objective is ease of search. In web architecture and development, new devices and innovations are what bring changes to dynamics of responsiveness. So, customization and client comfort are two driving forces for compatibility and responsiveness [31].

2.7.1 What Is Responsive Web Design?

The webpage and its design's acceptance, along with view as per screen size, platform, and direction of use, determine the responsiveness. Comfort in search and compatibility with the use give maximum boost to the success of the website [32]. The design of the website depends on *CSS* media questions. When a visitor changes from PC to mobile, then the view of the website should change accordingly and naturally respond to the inclination of view. Transplant this content order to the web plan, and we have a comparable yet entirely different thought. For what reason would it be a good idea for us to make a custom web plan? All things considered, don't developers make a structure for each client's required size and type that is associated with it? Like responsive engineering, web configuration ought to naturally change. It shouldn't require incalculable uniquely designed answers for each new visitor. Auto changes and sensor-induced responses are not what should be focused on. It is the dynamism that decides on what is prudent and paramount. Content, media, and designs all have to be synergistically articulated with a clarity required for responsiveness. Vertical scroll is what is mostly required for a better view. There are some highlights for responsive web design and viewpoint setting as given below:

1. Composition of websites for different devices and screen sizes and inclination of view.
2. Use of just HTML and CSS for resizing, covering up, subsiding, broadening, and moving the content.
3. Responsiveness is not a program or a JavaScript.

4. Focus is on the client's zone of interest of a website page.

5. The viewpoints of visitors always fluctuate from device to PC.

6. Before mobiles and phablets came along, the website was primarily constructed with a static plan and fixed size. After their advent, the downsizing pages came into picture which are now prevalent to fit the screen size.

7. HTML5 introduced a technique which let website developers take the responsibility to present by the *<meta>* tag.

8. Incorporation of the *<meta>* tag throughout the website pages is mainly done.

9. The programming is *<meta name="viewport" content="width=device-width, introductory scale=1.0">*.

10. A *<meta>* viewport component provides the requisite guidelines for measurement and scaling.

11. The *width=device-width part* sets the requisite width for the devices.

12. The *scale=1.0-part* sets zoom in and out for the layout view.

Some additional rules to follow:

1. *Don't use large fixed width*: Pictures and videos should not cover all the view width. It gives a not so responsive image of the website.

2. *Don't rely on content for a selected width to view*: CSS pixels shift between devices and they control the particular view.

3. *Utilize CSS media queries for different screen sizes*: CSS widths for pages and media must be used as the component for viewpoint on a device. It shouldn't place the media outside of the viewer's width of vision.

2.7.2 Advantages of Responsive Web Designs

- Content and pages become naturally adaptable, fluid, and give a uniform view to all the viewers across different devices and operating systems. User experience is only increasing with all the positive views.

- Financially also, responsiveness gives the website a significative position by helping out in wiping the extra cost across all platforms. SEO can be done well for more traffic as it is very simple along with its user friendliness [33].

2.8 Conclusion

This chapter describes the guidelines required for a compelling website, web traffic generation, maintenance, and responsiveness of sites. In this way, the web developer ought to consider a decent SEO and SEM alongside site improvement to get logical clicks for the website. Across all interfaces, if site is all around kept up, at that point it will give great traffic and the content can be completely improved. The different systems of SEO and SEM rely upon this rule of thumb that, without bending or any unsettling influence, if the content is managed effectively and efficiently, then subsequent pages will be positioned appropriately. Web index-controlled pages are the aftereffect of well-managed and represented content.

References

1. Ali, R., & Beg, M.S. (2011). An overview of web search evaluation methods. *Computers & Electrical Engineering*, 37(6), 835–848.
2. Beall, J. (2006). The death of metadata. *The Serials Librarian*, 51(2), 55–74.
3. Can, F., Nuray, R., & Sevdik, A.B. (2004). Automatic performance evaluation of web search engines. *Information Processing & Management*, 40(3), 495–514.
4. Croft, W.B., Metzler, D., & Strohman, T. (2010). *Search engines: Information retrieval in practice* (Vol. 520). Reading: Addison-Wesley.
5. Iredale, S., & Heinze, A. (2016, September). Ethics and professional intimacy within the search engine optimisation (SEO) industry. In *IFIP international conference on human choice and computers* (pp. 106–115). Springer, Cham.
6. Katumba, S., & Coetzee, S. (2017). Employing search engine optimization (SEO) techniques for improving the discovery of geospatial resources on the web. *ISPRS International Journal of Geo-Information*, 6(9), 284.
7. Krutil, J., Kudělka, M., & Snášel, V. (2012, November). Web page classification based on Schema. org collection. In *2012 fourth international conference on computational aspects of social networks (CASoN)* (pp. 356–360). IEEE.
8. Singh, S., Mondal, S., Singh, L.B., Sahoo, K.K., & Das, S. (2020). An empirical evidence study of consumer perception and socioeconomic profiles for digital stores in Vietnam. *Sustainability*, 12(5), 1716.
9. Singh, L.B., Mondal, S.R., & Das, S. (2020). Human resource practices & their observed significance for Indian SMEs. *Revista ESPACIOS*, 41(07). Retrieved from http://www.revistaespacios.com/a20v41n07/20410715.html
10. Sharma, E., & Das, S. (2020). Measuring impact of Indian ports on environment and effectiveness of remedial measures towards environmental pollution. *International Journal of Environment and Waste Management*, 25(3), 356–380. doi: 10.1504/IJEWM.2019.10021787
11. Das, S. (2020). Innovations in digital banking service brand equity and millennial consumerism. In Kamaljeet Sandhu (ed.), *Digital transformation and innovative services for business and learning* (pp. 62–79). Pennsylvania, PA: IGI Global.
12. Mondal, S.R. (2020). A systematic study for digital innovation in management education: An integrated approach towards problem-based learning in Vietnam. In Kamaljeet Sandhu (ed.), *Digital innovations for customer engagement, management, and organizational improvement* (pp. 104–120). Pennsylvania, PA: IGI Global.

13. Singh, S., & Das, S. (2018). Impact of post-merger and acquisition activities on the financial performance of banks: A study of Indian private sector and public sector banks. *Revista Espacios Magazine, 39*(26), 25.

14. Jain, S., Jain V., & Das, S. (2018). Relationship analysis between emotional intelligence and service quality with special evidences from Indian banking sector. *Revista ESPACIOS, 39*(33), 3–16.

15. Das, S., Mondal, S.R., Sahoo, K.K., Nayyar, A., & Musunuru, K. (2018). Study on impact of socioeconomic make up of Facebook users on purchasing behavior. *Revista Espacios, 39*, 28–42. Retrieved from http://www.revistaespacios.com/a18v39n33/18393328.html

16. Mondal, S., Das, S., Musunuru, K., & Dash, M. (2017). Study on the factors affecting customer purchase activity in retail stores by confirmatory factor analysis. *Revista Espacios, 38*, 30–55. Retrieved from http://www.revistaespacios.com/a17v38n61/17386130.html

17. Mondal, S., Mall, M., Mishra, U.S., & Sahoo, K. (2017). Investigating the factors affecting customer purchase activity in retail stores. *Revista ESPACIOS, 38*(57), 22–44.

18. KumarSahoo, K., & Mondal, S. (2016). An analysis of impact of electronic customer relationship management (e-CRM) on service quality of e-retail stores: A study of Bhubaneswar. *Research Revolution, Vol* (2), pp. 10-12. *INDEXED BY, 10.*

19. Mondal, S., & Sahoo, K.K. (2020). A study of green building prospects on sustainable management decision making. In Arun Solanki and Anand Nayyar (eds.), *Green building management and smart automation* (pp. 220–234). Pennsylvania, PA: IGI Global.

20. Das, S., & Nayyar, A. (2020). Effect of consumer green behavior perspective on green unwavering across various retail configurations. In Vannie Naidoo and Rahul Verma (eds.), *Green marketing as a positive driver toward business sustainability* (pp. 96–124). Pennsylvania, PA: IGI Global.

21. Das, S., Nayyar, A., & Singh, I. (2019). An assessment of forerunners for customer loyalty in the selected financial sector by SEM approach toward their effect on business. *Data Technologies and Applications, 53*(4), 546–561.

22. Subhankar, D., & Anand, N. (2019, May). Digital sustainability in social media innovation: A microscopic analysis of Instagram advertising & its demographic reflection for buying activity with R. In *1st International scientific conference "modern management trends and the digital economy: From regional development to global economic growth" (MTDE 2019)*. Atlantis Press.

23. Singh, I., Nayyar, A., Le, D.H., & Das, S. (2019). A conceptual analysis of internet banking users' perceptions. An Indian perceptive. *Revista ESPACIOS, 40*(14), 1–17.

24. Mohanty, P.C., Dash, M., Dash, M., & Das, S. (2019). A study on factors influencing training effectiveness. *Revista Espacios, 40*, 7–15. Retrieved from http://www.revistaespacios.com/a19v40n02/19400207.html

25. Singh, I., Nayyar, A., & Das, S. (2019). A study of antecedents of customer loyalty in banking & insurance sector and their impact on business performance. *Revista ESPACIOS, 40*(06), 11–28.

26. Gupta, D.K., Jena, D., Samantaray, A.K., & Das, S. (2019). HRD climate in selected public sector banks in India. *Revista ESPACIOS, 40*(11), 14–20.

27. Malaga, R.A. (2010). Search engine optimization—black and white hat approaches. In Marvin Zelkowitz (ed.), *Advances in computers* (Vol. 78, pp. 1–39). Cambridge, MA: Elsevier.

28. Nogueras-Iso, J., Zarazaga-Soria, F.J., Lacasta, J., Béjar, R., & Muro-Medrano, P.R. (2004). Metadata standard interoperability: application in the geographic information domain. *Computers, Environment and Urban Systems, 28*(6), 611–634.

29. Aalders, H.J. (2005). An introduction to metadata for geographic information. In Harold Moellering, H.J. Aalders, and Aaron Crane (eds.), *World spatial metadata standards* (pp. 3–27). Oxford, UK: Elsevier Science.

30. Zhang, J., & Dimitroff, A. (2004). Internet search engines' response to metadata Dublin core implementation. *Journal of Information Science, 30*(4), 310–320.

31. Onaifo, D., & Rasmussen, D. (2013). Increasing libraries' content findability on the web with search engine optimization. *Library Hi Tech, 31*(1), 87–108.

32. Zhang, J., & Dimitroff, A. (2005). The impact of webpage content characteristics on webpage visibility in search engine results (Part I). *Information Processing & Management, 41*(3), 665–690.

33. Nadanyiova, M., & Das, S. (2020). Millennials as a target segment of socially responsible communication within the business strategy. *Littera Scripta, 13*(1), 119–134. doi: 10.36708/Littera_Scripta2020/1/8

3

On-Page Optimization

3.1 On-Page Search Engine Optimization

This is the technique of improving individual webpages so as to rank higher and secure dynamically huge traffic by web crawlers. On-page implies both the available content and HTML source code of a page that can be developed, as opposed to off-page SEO which proposes links and related links.

3.2 Keyword Significance

Keywords are the basic and most important SEO tools for any web browser as they help the search queries to be materialized and actualized for search queries. So, choosing the right keyword is the main activity for redesigning SEO operation. If a web developer cannot design good keywords, then SEO will be very difficult to operate as a lot of money and time will go down the drain [1]. There are processes for choosing the right keywords and every SEO developer should use them.

3.2.1 Selection of the Right Keywords for SEO

For a dynamic web search, the visitor will always want to have a two-word- or three-word-based keyword by which the search queries can be easily pop up and the information can be accessed easily [2]. For example, for a website about canines, don't endeavor to plan for the keywords "pooch" or "mutts". Or maybe you could endeavor to focus on phrases like "dog passive consent getting ready", "minimal canine breeds", "specially designed pooch sustenance", "dog sustenance plans", etc. It is very difficult as well as inconvenient to have one or two word keywords and much of the time is not worth it, so, it is prudent to concentrate on less centered, significantly prominent keywords. The vital thing you need to do is to prepare keywords that portray the content of your site.

Ideally, the marketer knows the customers well. So, the search queries and keywords of potentiality can be more suitably suggested by the marketer to the developer and should be given priority. When picking the suitable keywords to update, you need to consider the ordinary month-to-month number of visits or queries as well as the congruity of these specific keywords to the website. Keywords show how much the website takes the point of view of visitors or the perception of the information search process [3].

3.2.2 Keyword Numbers in Website

Once you have found the keyword which depicts and represents the core theme of the website and expectedly gives the maximum result in attracting visitors, then it is the duty of the developer to make the website abundant with similar words. But excessive keywords will make the website cluttered. However, the optimized number of keywords will give 3–7% more importance than a lesser number. So, it is important to have good keywords in the content. It is good to keep five to ten keywords on a page so that it will be optimized well [4]. Moreover, what is progressively lamentable is that there are no easy ways (checking confinement from the web file) for specific keyword cluttering since this is seen as a deceptive practice that tries to control the questions which may arise later.

3.2.3 Keywords in Special Places

Keywords in optimized numbers increase the quality additionally. If keywords are in the page title, the headings, and the principle areas then it is good to have a link with underlying content. If you have a comparable special keyword as your competitors, yet you have that word in the URL, this will help the website fantastically in improving the rank [5].

1. *Keywords in URLs of Website*

    Make an effort not to overjoy when pursuing a qualitative keyword-rich URL. As per the rule of thumb of SEO, we should have five keywords in the URL of the site, as otherwise it will be difficult to remember them. So, there must be some balance in it. Additionally, pay a little attention toward whether you know how to consider a few extraordinary website names, which are already taken by others. So, a tool which helps you find a URL name will be valuable. File names and directory names are similarly noteworthy as they are easy and less complex to remember [6].

2. *Keywords in Webpage Titles*

    The webpage title is exceptionally accepted as the content of the <title> tag for the most part appears in most web data (searching in Google). Though it is not compulsory, for SEO purposes developers generally don't leave *<title>* empty.

3. *Keywords in Webpage Heading*

Normally headings of webpages are differentiated from subtitles from a dynamic website. It is not so important to have a heading after another entry, but from an SEO point of view, this is very much in practice. There is also no limitation with the length of headings as there is no restriction on page number. It is advisable to have at least seven to eight keywords in a heading, or it may spread to two to three lines which should be avoided. If we keep keywords continuously in two to three lines then it will not be good [7].

3.3 Domain

Website names are the most important thing for getting clicks on the Internet. They incorporate three fragments: A top-level domain (at times called an expansion or suffix), a domain name (or IP address), and a discretionary subdomain. In the URL http://www.xxxyyyzzz.com, we have *http://* which represents the protocol, *www.* denotes the subdomain, *xxxyyyzzz* is the domain name, and *.com* is the top-level domain. Both the domain name and top-level domain combined represent the root domain.

The top-level domain (TLD) appears for a website name.. Some instances of top-level domains include .com, .net, .org, and .edu. While we're likely all familiar with the TLDs above, there are, as a general rule, in excess of 1,000 possible TLDs from which site administrators can pick. TLDs are related with particular countries or areas like .uk (United Kingdom) or .dk (Denmark). These country (and now and again territory) unequivocal TLDs are known as country code top-level domains (ccTLDs).

3.4 Selection of Correct Domain

Before you select the correct domain, you should set aside some genuine effort to think about the domain name that you pick and, in all honesty, the brand name that you pick and how that is addressed through your domain name on the web [8]. Domain names have an enormous impact on dynamic browsing for getting more clicks and developing links.

3.4.1 Domain Name Should Be Brandable

Always the domain name should be such that it helps in creating a brand and identifiable features. Hyphens and numbers are a certifiable issue since they don't make something sound like a brand, nonexclusive or impossible to miss. For example, the new association that you are starting is a website that has pasta plans and conceivably sells some pasta-related web business things on it. Then here traditional keywords won't work as they are difficult to recall and remember [8]. You need something extraordinary. You should in all likelihood abstain from something like PastaRecipesOnline.com as it is unnecessarily nonexclusive, while BestPasta.com sounds to some degree better. PastaAficionado sounds impressively progressively brandable. However, it's verifiably novel. PastaLabs.com sounds brandable, unique, significant, and it stands out. You will recall it. It has kind of a sensible importance to it.

3.4.2 Make It Pronounceable

For what reason is it so critical that a domain name should be pronounceable? A large number of individual letters will form the name or they will tap on an association, so for what reason does it have any kind of effect?

It has any kind of effect by virtue of an idea called *"getting ready recognition"*. It's a mental inclination that individuals have where, fundamentally, we review and have continuously valuable associations with things that we can, without too much of a stretch, say and viably consider, and that represents a website. Regardless, totally, if you can only with huge exertion say the name, and others are not viably prepared to consider how to say that name, you will lose that ready recognition; you will lose that memorability and all of the upsides of the brand identity that you've made. For example: FlourEggsH20.com., Raviolibertine.com, LandofNoodles.com.

3.4.3 Make It Short

This implies that the domain name should be such that it will have minimum words and be identifiable. Short domain names make things look very simple and yet powerfully placed in the subconscious minds of consumers [9]. So, the rule is *keep it short and simple.*

Bias towards ".com": The web's been around for at least 20 years. For what reason does .com matter so much when there are countless TLDs available? The suitable answer is, yet again, that this is the most seen, most viably accessible brand outside of the tech world.

3.4.4 Don't Use Those Names Which Interfere with Other Similar Organizations

It is unlawful to use similar words or keywords or spell-alike words for your websites which are in direct competition with the competitors. One should avoid these names as they can attract legal disputes and a bad reputation in the market as the competitor can file a lawsuit [10].

3.4.5 Make It Intuitive

If your domain name is easily intuitive for all who easily interact with you, then it is better to have that name as your domain name. It can give you the advantage of quick reliability and relatability among the visitors [11]. With the option to look for a domain name and states as example, "Goodness, I'm perceiving what the visitors likely do. This is in all probability what that association is doing".

3.4.6 Always Go for Broad-Scope Keywords

This can help generally for mental commonality and planning recognition tendencies that we've talked about, yet, what's more, a touch from an SEO perspective because of the key messages that you generally will gather when people interface on the domain. Unless seen by Google as an exact match or fragmented match, you can use them.

For example, it would be better not to choose something like RecipesForPasta.com or something like BuyPastaOnline.com. You could go for something peculiar like Gusto.com. Consider a brand like Amazon.com, which indisputably has no association with what it is, or Google itself, Google.com. These are incredibly well documented and associated with their strengths, yet they don't generally require a potential keyword. If the domain name is not available then it is fine to append and modify, then search for it.

3.5 Planning and Designing Website Structure, Content, and Hosting

A website is one kind of personal communication tool which an organization uses for interacting with the world. It is a place of dynamism which contains information in the form of webpages and is unique in its own way as compared to other formats of communication [12].

3.5.1 Designing Process for Websites

Planning requires pertinent dynamic content as a write-up, magazine article, or noteworthy TV ads. Furthermore, planning for a website should be refreshed every now and again and remain in a state of harmony with evolving innovation and information as visitors depend on it for data. There are three stages of development involved in website development: Getting a domain name, finding a web administration process, and designing and formatting the website [13]. So, the designing process involves five steps:

1. Analysis of content towards target audience with all details.
2. Organizing content in layout and design for easy navigation.
3. Developing web layout, site structure, web page elements, and graphic designs.

4. Implementing visitor interaction, final checklist, and final touch to the content.
5. Maintaining supportive activities like SEO, analytics, and marketing.

3.5.2 Scope of the Website

First of all, the need analysis should happen, where the utility and benefit of the website has to be considered from the following points.

1. Promotion of IDEA, INTEREST, and INFORMATION
2. Advertise PRODUCT or COMPANY
3. Quick reach to a large AUDIENCE
4. Educate the VISITORS
5. Expand the BUSINESS
6. Provide DATA or INFORMATION

Before you can begin choosing what data the site will contain, you have to decide these following aspects:

1. Interest group for website
2. Time requirement for visitor
3. Internet knowledge of visitor and traffic
4. Effective communication and motivation for website idea

For preparing suitable optimized content one needs to keep the above objectives in mind, like the profile of the interest group, the visibility and viability of the information, details of data which visitors will browse, and the usability of pictures, videos, and graphics. The potential visitors and their profiling also need a thorough examination where their digital socioeconomic profile along with their psychographic views can be studied [14]. The objective of content always guides research on the visitors. Content can include the following items for more visibility:

1. Regular refreshed data
2. Articles about the products and organization
3. Frequently asked questions with their answers about the product
4. Digitally purchased items
5. Place to record who has visited and collect feedback
6. A section on the similarities and uniqueness of your website compared with others
7. Graphical representations

8. Weekly survey for browsers of a website
9. Quizzes, free offers, and related competitions for attracting visitors with prizes
10. Articles published in the public domain
11. Uniqueness of information
12. Maps for describing the particular location
13. Space for financial transactions

The most used practice for web designing includes (a) knowledge of visitors and traffic mobilization, (b) keeping webpages short and crisp, full of potential keywords, (c) limited word use for describing content, (d) not including huge images that block the memory space, (e) use of online-safe hues, (f) clarity in identification of all connections, (g) good spell check, (h) good use of webpage content, (i) proper updating information, and (j) inclusion of authentic reliable data [15]. New developers can always go about their actions by examining the scheme, layout, navigation, content, and presentation techniques of established websites [16].

3.5.3 Overall Designing of Website Structure

Overall design includes theme, navigation, and representation of content. There are three different navigation methods: Linear, database, and hierarchical navigation, which help in all the movements in the content (Figure 3.1).

3.5.4 Linear Navigation

The direct route is more convenient for a visitor for moving between various steps. This is normally utilized inside a site, however sometimes as an independent structure. Here visitors go with a predefined

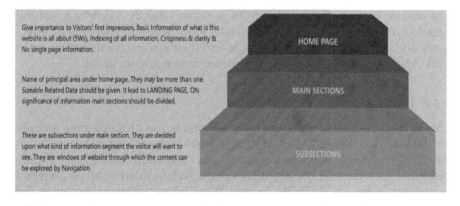

FIGURE 3.1
Website structure. Source: Author's conception.

STRAIGHT LINE LINEAR NAVIGATION

FIGURE 3.2
Straight line linear navigation. Source: Author's conception.

arrangement of navigation set by the developer on the website which is generally seen in tutorials. Here direct sequential straight activities are navigated (Figure 3.2).

Program the connections with the goal that they just grant development linearly for the entire website. The connections are connected so as to constrain the guest to begin at one side and proceed to an end.

3.5.5 Linear Reciprocal Links

This allows the navigation to move back and forth between all the pages during browsing. It is an extension of linear navigation where reciprocation and interconnectivity matter between pages.

3.5.6 Database Navigation of Webpages

The web content database is here comprised of a matrix and divisions and their own structures. This will be helpful when a lot of information is to be given in the web architecture for navigation, like a census database.

3.5.7 Hierarchical Navigation of Webpages

The various leveled structures go from general to specific, i.e. from a landing page to further divisions and subdivisions. Here browsing can be done effectively without completing a whole page and then waiting for the other one to load. Also, you have the option of returning and going up again.

3.5.8 Hosting

This is the most important stage after arranging and building up a website for the Internet. Who will be facilitated? Would it be advisable for you to go for SEO? Web facilitation, or hosting as it is popularly called, isn't simply identified with the cost and highlights of the web facilitating server. Website proprietors who are distantly located think about different variables, for example, website design and subsequent improvement. Does picking a host server have any effect on SERP or not?

There is a solid logic for increasing the rank with a remarkable IP address. Then again, it doesn't appear to be so important nowadays as most sites are facilitated on shared IPs. On a solitary PC, numerous servers run one next to the other and serve various spaces from single IPs [17]. If one goes for SEO, then there are web facilitating worries. Server issues like hosting a similar type of website won't help all. That may lead to spamming, replication of websites, or it may attract websites for illegal activities such as gambling or pornographic content to be used with the websites, which reduces the cost and is 10 times lower per month for 5,000 MB and unlimited space; however, "does that influence my position and does it harm my site?" The most convincing response may be yes and no. It may or may not influence, but situations may affect ranking. If a virtual IP is used then there is a chance that Google may spam it or block the browsing [18]. If illicit content is there, or if the webpage is boycotted by Google, your site will be prohibited. While you ought to never accept that your site is boycotted, you ought to likewise play it safe.

For appropriate site administration and management, hosting must be up 100% secure and safe, so that visitors won't feel cheated. Essentially, search engines have no particular calendars for creeping, so hosting must be up consistently to make web crawlers work well. If Googlebot experiences a 404 error, it will come back again in the following crawl, so, to err on the side of caution, guarantee 100% uptime of your site [19]. Then again, if your host doesn't educate you and you see a 404 error, then you should think about moving to another server for hosting. Googlebot can check whenever it wants, so your site must be prepared to obtain proper, legal, and safe hosting [20].

3.5.9 Web Hosting and Clean Records

The past history of the host needs to be very good in Google search. Despite the fact that there is no instrument to rate them, web hosts that are mainstream and widely utilized could be trusted as opposed to local ones. Virtual shared IP is a framework by which a web server utilizes a similar IP address for various websites. Rather than having another PC for a server, numerous virtual servers are made to dwell on the same PC. Some web hosts can have 2,000 websites on a solitary IP address; everything relies upon the content and assets accessible to the server. For this situation the site page's feed to the web crawlers would be moderate. Anyway, nowadays, virtual shared facilitating is a good alternative, and for the most part acknowledged by Google. Google doesn't boycott sites because they share an IP address with a site that is acting improperly. In the expression of Craig Silverstein of Google,

> Really, Google handles for all intents and purposes facilitated websites and their connections simply equivalent to areas on one of a kind IP addresses. In the event that your ISP does virtual facilitating accurately,

you'll never observe a contrast between the two cases. We do see a little level of ISPs consistently that misconfigure their virtual facilitating,
which may represent this persevering misinterpretation.

This makes it very certain that Google doesn't punish a website because of
virtual facilitating if done properly [21].

3.6 URL Restructuring

For maintaining the correct structure of your URLs is likewise significant.
Search engines face issues while ordering in dynamic websites. *Apache* is
the most prevalent web server and modifies the module to change itself
to dynamic long URLs for search engines, and most facilitating organizations offer it. For instance, when it is introduced it can change to accompany
unique locators. Facilitating should not be a worry for your site. Contingent
on your experience and specialized expertise, you may profit by going with
the most famous hosting management, regardless of whether they are somewhat more costly, as opposed to going for the best arrangement of a host
with no notoriety. If Google has an issue with your host, at that point you
will in all probability have issues with Google. From numerous points of
view, it is desirable to have an individual IP address with no illegal activities
associated with it. Websites facilitated by free web hosts can't be relied upon
to rank highly on Google for the most potent keywords, as free websites
often use these words in their content. More or less you should be very interactive with your web server and explain all issues in detail before selecting
the host. These little contemplations can give you a sense of security for all
the future modifications and activities for developing the website [22].

3.7 HTML and XML Sitemap

These techniques are vital for landing page SEO. When you are finished
making your sitemap, it is prescribed that you submit it with Google webmaster tool with the goal that Google can crawl your sitemaps. In any case,
individuals frequently get confounded when it comes time to choose which
sort of sitemap is smarter to utilize. Well, everything relies upon your site's
interest. One of the primary contrasts between HTML and XML sitemap, is
in how Google views the site [23].

3.7.1 The Rundown

Before stepping into the details, we should simply understand that XML
is meant for search engines, while HTML is beneficial for users. The XML

sitemap convention is explicitly planned for special search engine spiders. At its root, XML is a record that incorporates all the background movement on a site. The site's fundamental URL along with all URLs constitute the metadata. This can incorporate when the URL was last refreshed, how significant it is, the normal repeated changes happen, the URL's connection to the remainder of the site, and so forth. HTML is only a general review of the site, only the pages and information that one should be worried about. If you're on a site and you're searching for the shopping basket or the "Get in Touch with Us" page and can't discover it, you'd go to the sitemap and effectively discover it there. While this is not suitable for a browser, it helps your search engine positioning in light of the fact that your website is easy to use and taking into account the webpage guest [24]. XML increases the possibilities of search and browse for the site. If the site is capable enough to generate a good search and rank, then it may not be necessary. However, the XML sitemaps truly help web search tools crawl through your website in a considerably more productive manner; they alert search engines to your quality and ensure you are getting the majority of your significant pages viewed. The HTML webpage guide is connected to from each and every page of your site. It is additionally utilized for visitor experience so when you compose this kind of program you need to consider the visitor. We need a HTML sitemap to give our visitors a one of a kind viewing experience by developing visitor direction. Visitor direction is critical in light of the fact that they need to know where these sitemaps are on your site in connection to everything else. So, if a visitor goes into your HTML sitemap, you need to ensure the majority of your significant classifications of subpages are considered. Nonetheless, you don't need each page to be linked with the HTML sitemap like on an XML sitemap, since that would not bode well for a good view. Contingent upon the size of the site, we by and large need to have close to between 100 and 200 pages on the XML sitemap page. For small websites, 50 pages will be sufficient. Having both HTML and XML sitemaps will keep your website sure and secure enough to ensure that you are not passing up any part of SEO. It will guarantee that Google can discover web pages and good traffic can be generated. Having both sitemaps is the most ideal approach in the event that you need your site both positioned and easy to use.

3.8 URL Structuring

Web optimization and well-disposed URLs have always been in focus for SEO. From the quantity of organizers to the consideration of the essential specific keywords, proficient SEOs have investigated various mixed approaches for the ideal URL structure. It is described below how to structure or formulate URLs.

3.8.1 Keywords Inclusion

Keywords inclusion is something to be thankful for. In the event that you are attempting to streamline a blog entry for a specific keyword, you ought to incorporate that fundamental catch phrase in its URL. If the fundamental keyword is included naturally with the content, then it will draw good response. Benefits are as follows:

1. It will come up in the status bar if a browser clicks on a linking website that comes up as a picture.
2. Another huge preferred method is that web crawler visitors give a great deal of consideration to significant catch words in the web index SERP which draws greater traffic.

3.8.2 Giving Proper Reading to the Content Words

The most effective URL is that which the visitors can use easily. Web index is the not the most important thing for a clean URL, since visitors generally go for content and click on long items to read. Web index reads solid true information and informs their visitors to recognize if a positioned page is deserving of its internet search position. What's more, a progressively meaningful keyword can fundamentally develop the clicks. The idea is that the URL ought to be effectively recognizable [25].

3.8.3 Keep It Short and Simple

Web search tools don't especially have any issue with longer URLs. In this way, at the end of the day, URL length doesn't directly influence the web index and rankings. Nonetheless, here only the visitor experience and visitors' views about comfort of browsing are the issues which one will look for.

Short URLs are generally preferred as they are simple to use. Besides, due to the limitations of social media like Twitter, shorter URLs are much simpler to share over the internet. As a rule of thumb, if the URL of your blog entry is under 50 characters, you don't need to stress over it. If it is longer than 100 characters, at that point you should rework it and make it easier to understand. The URLs of articles positioned on Google's #1 page contain an average of 37 characters.

Aside from that, there is additionally an issue of incorporating classifications and subcategories in the URL structure. Mostly, you may realize this as including "organizers" in the URL of a post like www.xyz.com/categori es/mobiles/samsung-mobiles/top-10-mobiles-in-vietnam.html. Here, URL structure turns out to be excessively long in the event that you incorporate such a large number of categories.

Presently, the question is, does it help crawlers? Tragically, it doesn't generally help in improving a web crawler positioning. Be that as it may, simultaneously, such extensive URLs don't have all the symptoms of being very easy to understand. Famous websites explicitly utilize shorter URLs with no additional indexes.

3.8.4 Matching of URLs to Titles of Webpage

It is viewed as an excellent practice to coordinate the URL with the title of the post – in any way possible. This is on the grounds that when somebody opens the link to your blog entry in the wake of seeing its URL and the keywords in it, he/she hopes to find out about a specific point [26]. If title or content of the page is to some degree not quite the same as the catch words present in the URL, it might prompt a "bounce". At the end of the day, when web index visitors open your blog entry, they anticipate something. You must meet those desires and convey the correct content.

Consider the possibility that the title or feature of your blog entry has stop words in it. In a perfect world, you should coordinate the title and URL as much as possible. So, what befalls each one of those stop words? Luckily, you don't need to stress over that.

Stop words, like *and, or, at the same time, the, an,* and so on, are not important to incorporate into the URL. There is no restriction for inclusion of these words in the URL; however, they can frequently make a URL irrationally long. As you can recollect, one of our objectives is to consistently make the URL short, basic, and decipherable. Keeping away from such stop words can enable you to accomplish that objective of effortlessness [27].

3.8.5 Case Sensitivity

Not a great deal of bloggers and site proprietors focus on it, yet case affectability can be a major issue. If you are utilizing the Microsoft/IIS servers, you don't need to stress over it by any stretch of the imagination. Be that as it may, in the event that you are facilitating with Linux/UNIX, there can be a bit of an issue. This is on the grounds that facilitating with Linux/UNIX implies that bloggers can decipher separate cases. For example, all things considered, www.abc.com/xyz might be unique in relation to www.abc.com/XYZ. As should be obvious, there are no enchantment runs here. The greater part of the advice is centered around making the easiest to use and comprehensible URL for your website pages. Aside from that, including the essential keywords and after that coordinating sensitivity with the title and content of the post are other significant elements that you should concentrate on. Be that as it may, in a way to incorporate those keywords, it is significant not to over-streamline the case sensitivity as it can prompt a Google block. Web search tools have experienced a monstrous update in the previous couple of years, and now they are more intelligent than at any time in recent memory. The over-enhancement of keywords doesn't especially work any longer – regardless of whether it is in the content of the post or it is in its URL. Concentrate on making superb content, so Google and other search engines recognize it in their SERPs. At that point, make an SEO-accommodating URL structure by utilizing the tips referenced above to supplement it. That ought to be the center of your SEO system.

3.9 Picture Naming and Concepts of Title, Alt Tags

Well-placed and neatly presented pictures attract the visitors to navigate through the content. The alt traits of these pictures should not be neglected as alt tags and title tags fortify the message for web index. This improves the ranking. The expression "alt tag" is generally utilized for restricting what's really an alt trait on an img tag. For any picture on the website, the alt tag ought to depict what's on it. Screen followers for the visually impaired and outwardly weakened will not browse this content and along these lines make your picture available. When a picture is used without any logic only to create a background, then it ought to be in your CSS and not in your HTML. If you truly can't transform it, give it a vacant alt characteristic, as so: *<img src="image.png" alt="">*. The unfilled alt trait ensures that screen browsers skirt the picture, alt content, and SEO. Google, in their article about pictures, has a heading "Make extraordinary alt content". This isn't a suitable case; Google gives generally high esteem to alt writings to figure out what is in the picture, yet in addition to decide the subject of the encompassing content. This is the reason why, for instance, in the well-known SEO module, Yoast SEO content investigation, they have a check explicitly to see whether you have at any rate one picture with an alt label that contains your center keyword. Yoast SEO has the accompanying three phases for pictures and their alt message in your posts:

a. Red marked dot for *No images appear in this page, consider adding some as appropriate.*
b. Orange marked dot for *the images on this page don't have alt tags containing your keyword or phrase.*
c. Green marked dot for *the images on this page contain alt tags with the target keyword or phrase.*

This doesn't mean that you will spam the keyword into each alt tag. The only thing that is required is to have a calibrated a good related image for the posts where the website will have all the content around the centered keyword. If it is for a particular item, then one needs to include the ID and alt tag for effective search. So, the rule of thumb is that if a keyword in the picture helps in identifying something, then include the alt tag.

When you transfer the image to WordPress, then you can set the title and alt trait by using the picture file name. It simply duplicates the alt property. The appropriate alt content for each image should be added. For SEO Picture, it begins with the correct document name for the principal keyword.

3.10 Meta Tags

These are an incredible route for webmasters to give data about the websites to search engines. They give data for all browsers in all frameworks that they comprehend. They are added *to the <head> area of your HTML page.*

3.10.1 Job of Meta Tags or Tags in SEO

Positioning highly in Google in 2019 has unmistakably more to do with pertinence and notoriety of excellent content, visitor fulfillment, and fame than meta label improvement. Meta tags don't perceptibly impact where a page is positioned in Google in a positive way. Meta tags, when utilized appropriately, can in any case be helpful in various zones outside simply positioning pages. Misuse them, and you may fall foul of Google's correctional quality calculations [28].

3.10.2 Google Uses Meta Tags for Ranking

Some web crawlers once hoped to use concealed HTML tags like meta tags for requesting search engines in internet search results pages; however most search engines (in 2019) have developed past this, and Google surely has. Google has gone on record saying that it doesn't utilize few meta tags when positioning a page (in a positive way) and tests during the time have unquestionably appeared to affirm this.

3.10.3 Role of Meta Tags in SEO

Meta tags can help depict any page in an increasingly advantageous machine intelligible arrangement, progressively fit to web indexes, yet they are probably going to get spammed, thus gets limited, with regards to positioning archives on the web. Almost certainly, Google would search for wrong treatment in such tags and take action on it here and there, as opposed to remunerating it. Google may utilize metadata, among numerous different signs, to CLASSIFY pages, or DISPLAY data about a page in SERPs.

3.10.4 The Most Effective Method of Writing Meta Tags

You can even now be imaginative when considering a few tags like the meta description, yet meta tags are best utilized when *exactly* depicting the page being referred to, and helping Google to alternately process data about the page. If you are helping Google to help better particularly educational questions, and generating alternative ways to get information, then Google is your companion, and most can get profit by that relationship. At the point when utilized appropriately, meta descriptions customarily help structure the ultra-significant "snare" of your promoting in the free SERPs. Your tags ought to be exact, important, and distinct, and be cautious in concentrating them all pointlessly on only one keyword; however don't think "meta tags" until you have considered what the "subject" and "idea" of a page are, its "motivation", and end "user experience".

Fulfilling every requirement of a website is critical to building notoriety, in Google, and on social channels. If you are developing with Google, don't think page or article, think "point" or "subject" or data "center point". Google expects to return rich, enlightening article pages in *lots* of natural outcomes in future – that pattern is apparent in certain specialties. All real search engines suggest reasonable utilization of meta information, and in case you're composing helpful, unmistakable tags it will be far-fetched that any real web index will punish legitimate use – when no low-quality easy routes are taken, in any event. Most web crawlers use or have utilized meta data here and there to help characterize a record, however, if a web crawler "utilizes" meta-portrayal tags, for example, this doesn't mean they are utilizing it as a positive positioning sign of where your page positions in the SERPs. We will concentrate on the three most basic meta tags, *(a) description, (b) keywords, (c) robots*.

3.11 Keyword Generation Tools

The keyword search and generation process, regardless of whether for paid SEM or SEO, ought to consistently begin with three stages:

1. *Comprehend objective*: Search as per the objective of the website.
2. *Brainstorming*: Generate arrangements of the same number of applicable potential objective keywords.
3. *Analyzing*: Collect information and analyze the keywords on their potential. Investigate some free keyword revelation instruments and strategies you ought to utilize.

3.11.1 Google AdWords Keyword Planner Tool

It is the most sought-after tool for keyword setting and SEO. This tool is magnificent, yet not impeccable. The following are a couple of downsides that regularly should be tended to with different devices:

a. AdWords restores a great deal of insignificant keywords you shouldn't think about.

b. "Chunky center" keywords are frequently mysteriously hidden from rundowns.

c. Big keywords with under five to ten inquiries per month are totally rejected from the rundown of thoughts.

d. There is no real method to approve big keywords as something somebody may really scan for.

e. The reason for this tool is to build AdWords income, as it bodes well that the aftereffects of Keyword Planner steer visitors towards keywords to get maximum clicks.

3.11.2 Tips

Here are a couple of quick tips on bypassing superfluous outcomes and removing applicable long-tail keyword-related thoughts:

1. Entering the landing page or another site's URL can be beneficial.
2. Check AdWords and Keyword's Planner tab.
3. Addition of negative keywords.
4. Turn on the visibility of keywords planned with the website so that the main keywords can be searched.
5. Enter at least two words on one line to see keyword thoughts that incorporate every one of your words in any request.

3.11.3 Autocomplete-Based Keyword Suggestion Tools

There are a huge number of keywords that can automatically return every one of the outcomes from Google Autocomplete. Autocomplete tools are an extraordinary method to quickly go to the desired word trail from selected terms. They additionally sidestep AdWords' inclination for getting the visitors scanning for the longtail keywords that AdWords won't demonstrate. Google reported that it would cut off access to the Autocomplete API. Instruments totally dependent on this API would lose usefulness if access were cut. Be that as it may, numerous Autocomplete-based keyword recommendation tools are as yet working, as they can be fueled in different ways, for example, *Microsoft Bing Autocomplete* (another web crawlers' autocomplete) – it's not Google, yet despite everything it gives you a ton of extremely good actions like:

1. **SERP scrap**: There are instruments that can automatically carry out web crawler searches and constantly concentrate on autocomplete. The drawback is this disregards the terms of administration of all

web search tools: The crawler attempts to anticipate the word for saving money on their data transfer capacity, frequently blocking IP delivers that seem, by all accounts, to be scratching web crawler results pages. So, to control this they can hinder the click occurrence.

2. **Old historical information**: It is plausible for a keyword proposal device to store Autocomplete information and return that information when visitors utilize their tool.

3. **Selective API consent**: Here, Google allows Autocomplete API access to a few admins.

3.11.4 Quick Tips for Special Cases and Questions

Google Autocomplete can propose inquiries that start with your key term, yet in addition end with the key term, or are sandwiched in or around the key term. Before clicking "Enter", essentially type an underscore (_) after you type in the key term. Autocomplete is likewise an incredible method for creating word enquiries that the individuals try to search. Query-based keywords are great markers of what data visitors see. Also, keywords are suggested for frequently searched terms directly in cell phones [29]. Here are some Autocomplete-based keyword proposal tools:

1. Uber-Suggest is the keyword proposal for acting as a unique term-generator tool.. It appears to have changed to be fueled by Bing Autocomplete.

2. Soovle returns up to ten aftereffects of proposals from seven search engines as you type in the seed term. The top-ten result averts truly difficult work; however Soovle can be a fun method for commencing conceptualization.

3. ScrapeBox is an amazing keyword finding program for Keyword Harvester that can scratch the proposals of various seed terms from the SERPs of any web index at the same time.

4. AnswerThePublic has some clever representations of its keyword proposals which tries to get keywords from own data Continuously ensure you've completely inspected keyword information from your own site.

3.11.5 Search Terms/Queries for AdWords

If you are spending on AdWords, you have some entirely significant information for SEO keyword inquiry about. Then, recover keyword from the AdWords Search terms report or the Google Analytics AdWords Search Queries report. (Try not to be relaxed for generating "keywords", which are not the precise questions visitors scanned for.) Pay exceptional regard to inquiries with conversions.

3.11.6 Site Search

If your site has an internal site search, the terms visitors looked for might give you some content and keyword planning thoughts. Simply look at the Google Analytics Site Search Report. If you have not arranged site search, you may even now have the option to get some recorded information from page reports if your site searches have URLs in the organization string + search term, for example example.com/search/example+search+terms.

3.11.7 Other Methods for Collecting Keyword Ideas

General Q&A blog sites, such as Quora, can provide a decent form of applicable inquiries. Specialty discussions, similar to the ones at Moz or Houzz for instance, are amazing wellsprings of inquiries. Google Advanced Search can enable you to store information viably. Some of important Keyword idea generators are as follows:

1. **Antiquated search** on a point consistently provides a few thoughts too, so don't be hesitant to get all journalistic like searching on Wikipedia.
2. **Contender sites** can give some extraordinary thoughts. You can likewise utilize a potential keyword density tool to separate regular terms from a page, and you can utilize AdWords Keyword Planner to create keyword thoughts from some random page.
3. **Bing Webmaster Tool** is a slimmed-down or small variant of AdWords Keyword Planner that will create keyword thoughts and provides what Bing can look for.
4. **Rank Tracker** from SEO PowerSuite is keyword insight program that can pull in keyword recommendations from an incredible 20 sources (just each source in turn, however).

A thesaurus, as thesaurus.com or merriam-webster.com, normally demonstrates its value too.

3.11.8 Joining Two or More Lists of Terms into a Long Keyword

This system is an unprecedented strategy to find extra-long-tail keywords you may have missed. In like manner, it works best after you've perceived some uncommon strong keywords. There are many instruments to empower you to scale this astonishing method, including:

1. The Simple Long-Tail Keyword Generator is a mind-boggling spot to start joining keywords.
2. AdWords Keyword Planner has a worked in feature to join two records.

3. SEO Books' Keyword List Generator is an astonishing instrument that stores up to five records and allows you to connote match types and offers to help plan PPC keyword records.

4. Internet Marketing Ninjas Search Combination Tool is a keyword blend device that gives results that are intuitive interfaces with the SERPs for that keyword.

5. Excel Concatenation can be used to join terms in any way your heart desires.

3.12 Keyword Density Analysis

At whatever point you look for something in Google, you just observe what positions. You don't see the hundreds or thousands of pages that have been sifted for pushing excessively hard. In case you are uninformed on what density levels are sensible, consider designing your methodology based on what is working at the present time.

1. Search for your objective keyword in Google.

2. Grab five of the top positioned pages from the list items.

3. Analyze every one of them in a different tab utilizing a density examination device.

Remember that some exceptionally trusted brands rank more dependent on their image quality than the on-page content, in this way. If you are making content for a newer and less trusted site you would probably be in an ideal situation to put more weight on results from smaller and lesser known sites which have still figured out how to rank well on Google.

3.12.1 Keyword Density

There is no single perfect or comprehensive keyword density rate. Each search question is surprising, and web crawlers compare (or institutional-ize) reports against other top records to choose specific terms. A couple of keywords like "charge cards" typically appear as a two-word state, however various terms may be logically spread out. Further, some incredibly trusted locales, with unprecedented care, strong use data, and healthy association profiles, can very likely draw off more emphasis than tinier, less accepted goals for completing the keyword set up. If all else fails set up rules-of-thumb, concerning keyword repetition:

1. From a trusted corpus of internal content (like someone's inward site search, or a database of select acknowledged trusted content design-ers), higher is normally better.

2. From a wide corpus of external content (like general web search, where various people have a persuading power to endeavor to game the system), less is normally better.

3.12.2 Google On-Page Classifiers

At the point when Google revealed the primary Penguin update in April of 2012, they additionally revealed some on-page classifiers which punished a few pages that had over the top word redundancy. Apathetic and ignorant, modest, re-appropriated composing will in general be genuinely tedious – to a limited extent since individuals paid by the word to produce modest content have a motivating force to swell the word count, no impetus to cut back the excess, and no motivator to do profound research. Google's released remote rater rules advise raters to give a low rating to uninformed tedious content. Nowadays the essential utilization of these sorts of examination tools isn't to keep dialing up the keyword density, but instead to bring down the emphasis on the center terms while including exchange word structures, acronyms, equivalent words, and other supporting jargon.

1. **High density**: The upside of forceful redundancy, regarding helping to lift rank for the center term, is genuinely negligible, and high keyword density improves the probability that the page may get sifted.
2. **Low density (with variety)**: The upside of more noteworthy word variety (as far as helping lift rank for a wide assortment of related words) is critical, and lower density of the center terms diminishes the danger of the page getting divided.

3.12.3 Great versus Optimal versus Overdoing Keyword Density

Because of web spam, keyword density without anyone else's input is a genuinely poor indicator of importance. Restrictions constrained early web crawlers like Infoseek to depend vigorously on page titles and other on-page report scoring for importance scoring. In the course of recent years, search engines have become undeniably progressively ground-breaking. That has enabled them to fuse extra information into their importance scoring calculations. Google's enormous algorithms for getting down link information actually help other positioning elements as follows:

1. Search engines may put critical load on domain age, site authority like anchor text, localization, and usage of information in the form of data.
2. Each web search tool has its very own weighting calculations. These are diverse for each significant web index.
3. Each web crawler has its very own jargon framework which encourages them comprehend related words.

4. Some may put more weight on the above area-wide and offsite factors, while others may put more weight on on-page content.

5. The page title is conventionally weighted more than most of the other content on the page.

6. The meta keywords names, comments marks, and other hidden information sources may be given less weight than page copy. For instance, most colossal-scale hypertext web search tools put zero burden on the meta keyword tag.

7. Page copy which is bolded, associated, or in a heading tag is likely given more unmistakable weighting than commonplace content.

8. Weights are relative.

9. If your entire page is in a H1 label that looks somewhat off, and it doesn't put more weight on any of the content since all of the page duplicate is in it.

10. You presumably need to abstain from doing things like bolding the H1 message as it is suspicious; it will cause a page to appear to be progressively pertinent.

11. Excessive spotlight on density misses the mark on various fronts.

12. When individuals center a lot around density, they regularly compose content which individuals would not be keen on browsing or connecting to.

13. Some of the questions are somewhat irregular in nature. Generally, 20% to 25% of visitors' questions are one of a kind. At the point when website admins change up page duplicate for a discretionarily higher density, they commonly wind up evacuating a portion of the modifier terms that were helping the page seem significant for some three-, four-, five-, and six-word search inquiries.

14. Similar related algorithms may examine supporting jargon when deciding on the importance of a page. If you removed the keyword expression you were focusing on from your page duplicate, would it regardless be simple for a web index to scientifically show what that expression was and what your page is about given the supporting content? Provided that this is true, at that point your rankings will be undeniably increasingly steady *and* you will probably rank for a far more extensive crate of related keywords.

3.12.4 Would It Be Advisable to Incorporate Even Use Density Analysis Software?

These sorts of devices are still very important when utilized with the correct techniques. Utilizing investigation instruments can even now enable you to reveal a great deal of chances, including:

1. Looking at contending sites and finding some great expressions (and modifiers) to use in your page content, which you might not have seen at a careless look.

2. Helping you to check whether a page is way out of sync with top positioned pages.

3. Helping you decide whether an author is composing normally or utilizing over the top reiteration.

4. Your top rivals have been using their promotional tools for a considerable length of time.

5. Presently you can know precisely where they rank, pick off their best keywords, and track new open doors as they rise.

3.13 Creator Meta Tag Description and Uses

The creator meta tag is a meta label used to characterize the creator of the content on the site page. This meta tag enables you to credit the writer(s) of the content on the page. This meta tag is regularly utilized by CMS devices to keep up the responsibility for parts of a site [30]. It is particularly valuable when you have various content writers on the site. Another utilization of this meta tag is to detect counterfeiters or plagiarism. Regularly when individuals duplicate site content, they duplicate the whole page, including the meta tags [31]. By looking for content recorded as composed by you, you can demonstrate that they took your work all the more effectively.

3.13.1 Meta Tag Type

The creator meta tag is a named meta tag. Company name *<meta name="author" content="Author name(s)">*. You can place any content in the creator meta tag; however, it's ideal to utilize the full first and last names of the creator. In the event that there are numerous creators for a page, separate them by commas.

3.13.2 Prescribed Uses

The author meta tag isn't utilized by any web crawlers or programs; however, it might give extra data to web search tools that utilize the component content to depict the page. This is a discretionary label that you should utilize if your page requires it; however it won't include a great deal of significant worth. In the event that you are attempting to streamline your page for speed, forgetting about this meta tag is a smart idea, as it includes more content without including any worth.

3.14 Other Basic Meta Tags

Meta tags are components utilized on each page of a site to give data identified with the site and the individual pages of the site. Here are some basic meta tags:

1. <title>The Title of the Page</title>
2. <meta name="description" content="Description of page"/>
3. <meta name="keywords" content="Keywords identified with page"/>

3.14.1 Discretionary or Optional Tags

1. <meta name="geo. position" content="latitude; longitude">
2. <meta name="geo. place name" content="Place Name"
3. <meta name="geo. region" content="Country Subdivision Code">
4. <meta name="robots" content="noindex, nofollow"/>
5. <meta name="revisit-after" content="period">

3.14.2 Detailed Description

1. The title tag is for use on each page on the site. This ought to portray the content of every one-of-a-kind page and incorporate keywords if relevant.
2. The meta description tag is where you can incorporate a synopsis of the page content.
3. The meta keyword tag incorporates keywords significant to your site and the particular page.
4. The geotags are area-specific tags which can sit on each page on your site. These are for use if you have a physical area that is significant for your business, for instance a café. They can likewise be utilized for organizations that have stores all through the nation, with geotags doled out to any similar page for every individual area.
5. The robots label advises the search engine crawlers on how to treat the pages of your site; Noindex will advise the web crawlers not to record that page. Noindex will let them know not to pursue any of the connections on the page.
6. The revisit after label advises the web crawlers how frequently to return to and re-creep the site.

3.15 Accepted Best Practices for Title, Description, and Keywords

The title tag is significant and ought to show up on each page of the site. This ought to mirror the content of the page and incorporate a couple of keywords important to the page. The search engines find the keywords inside the title and description that matches with the result. Each page ought to have individual title tags.

The meta description ought to be expressive and furthermore identify with what shows up on-page. It must contain keywords where important and be confined to 155 characters (any extra characters will be cut off inside the web crawler results). The meta description has no/constrained positioning impact inside the outcomes, and in this way its activity is absolutely to get a click.

The meta keyword tag normally incorporates a couple of keywords significant to each page. This tag is presently to a great extent repetitive and is probably going to be overlooked by the web crawlers. In the early long periods of SEO, this tag was to a great extent mishandled by stuffing in loads of keywords, now and then inconsequential to the page content. To forestall this training, web search tools started disregarding these to counteract any control of the rankings. Whenever favored you can add a couple of keywords applicable to each page; on the other hand these can be forgotten about entirely.

3.15.1 Geo Tags

Geotags are commonly utilized if your site/business is area-specific. A café would be a genuine case of this as it has a physical area pertinent to the administrations it is advertising. These are area-specific and can be created utilizing different online devices (http://www.geo-tag.de/generator/en.html). Set on each page of the site, these let the search engines realize where you're based and could help improve your web search tool rankings for nearby related pursuit terms.

3.15.2 Robots Meta Tag

The robots meta tag is utilized to advise the search engines on how to treat certain pages of your site. For instance, in the event that you don't need the web search tools to record a specific page of your site, you can utilize the Noindex tag. If you don't need the web search tools to pursue any of the connections on a particular page of your site, you can utilize the Nofollow tag. It is suggested that you include these cautiously as they could avert the indexation of significant pages of your site whenever introduced on an inappropriate page or over the site.

3.15.3 Return to After Tag

The revisit-after tag is utilized to tell web search tools how regularly they should visit and search your site. A blog commonly is refreshed normally and would profit by being searched constantly by all tools of search. It is smarter to evade the utilization of this tag and leave it to the web search tools to choose how regularly to search your site. Current web search tools are truly exceptional and are probably going to have the option to separate a static site from a blog and will naturally search a site regularly for new content posted prominently.

3.16 Conclusion

Here one can conclude that on-page SEO works with the sitemap, page rank, and Keyword Planner coordinately and coherently. On-page SEO is the most utilized system of website development where the engineer keeps tabs on the content through online and different labeling procedures. Procedures are productive when executed appropriately with an obvious result and usage. So, this section sets up a detailed, documented method of on-page SEO and its subsidiaries.

References

1. Kumar, G., Duhan, N., & Sharma, A.K. (2011, September). Page ranking based on number of visits of links of web page. In *2011 2nd international conference on computer and communication technology (ICCCT-2011)* (pp. 11–14). IEEE.
2. Dubey, H., & Roy, B.N. (2011). An improved page rank algorithm based on optimized normalization technique. *International Journal of Computer Science and Information Technologies, 2*(5), 2183–2188.
3. Das, S., Mondal, S.R., Sahoo, K.K., Nayyar, A., & Musunuru, K. (2018). Study on impact of socioeconomic make up of Facebook users on purchasing behavior. *Revista Espacios, 39*, 28–42. Retrieved from http://www.revistaespacios.com/a18v39n33/18393328.html
4. Mondal, S., Das, S., Musunuru, K., & Dash, M. (2017). Study on the factors affecting customer purchase activity in retail stores by confirmatory factor analysis. *Revista Espacios, 38*, 30–55. Retrieved from http://www.revistaespacios.com/a17v38n61/17386130.html

5. Mondal, S., Mall, M., Mishra, U.S., & Sahoo, K. (2017). Investigating the factors affecting customer purchase activity in retail stores. *Revista ESPACIOS, 38*(57), 22–44.

6. KumarSahoo, K., & Mondal, S. (2016). An analysis of impact of electronic customer relationship management (e-CRM) on service quality of e-retail stores: A study of Bhubaneswar. *Research Revolution, 2,* 10–12.

7. Mondal, S., & Sahoo, K.K. (2020). A study of green building prospects on sustainable management decision making. In Arun Solanki and Anand Nayyar (eds.), *Green building management and smart automation* (pp. 220–234). Pennsylvania, PA: IGI Global.

8. Das, S., & Nayyar, A. (2020). Effect of consumer green behavior perspective on green unwavering across various retail configurations. In Vannie Naidoo and Rahul Verma (eds.), *Green marketing as a positive driver toward business sustainability* (pp. 96–124). Pennsylvania, PA: IGI Global.

9. Rani, P., & Singh, E.S. (2013). An offline SEO (Search Engine Optimization) based algorithm to calculate web page rank according to different parameters. *International Journal of Computers & Technology, 9*(1), 926–931.

10. Kumar, R., & Saini, S. (2011). A study on SEO monitoring system based on corporate website development. *International Journal of Computer Science, Engineering and Information Technology (IJCSEIT), 1*(2), 42–49.

11. Singh, S., Mondal, S., Singh, L.B., Sahoo, K.K., & Das, S. (2020). An empirical evidence study of consumer perception and socioeconomic profiles for digital stores in Vietnam. *Sustainability, 12*(5), 1716.

12. Singh, L.B., Mondal, S.R., & Das, S. (2020). Human resource practices & their observed significance for Indian SMEs. *Revista ESPACIOS, 41*(07). Retrieved from http://www.revistaespacios.com/a20v41n07/20410715.html

13. Sharma, E., & Das, S. (2020). Measuring impact of Indian ports on environment and effectiveness of remedial measures towards environmental pollution. *International Journal of Environment and Waste Management, 25*(3), 356–380. doi: 10.1504/IJEWM.2019.10021787

14. Das, S. (2020). Innovations in digital banking service brand equity and millennial consumerism. In Kamaljeet Sandhu (ed.), *Digital transformation and innovative services for business and learning* (pp. 62–79). Pennsylvania, PA: IGI Global.

15. Mondal, S.R. (2020). A systematic study for digital innovation in management education: An integrated approach towards problem-based learning in Vietnam. In Kamaljeet Sandhu (ed.), *Digital innovations for customer engagement, management, and organizational improvement* (pp. 104–120). Pennsylvania, PA: IGI Global.

16. Das, S., Nayyar, A., & Singh, I. (2019). An assessment of forerunners for customer loyalty in the selected financial sector by SEM approach toward their effect on business. *Data Technologies and Applications, 53*(4), 546–561.

17. Subhankar, D., & Anand, N. (2019, May). Digital sustainability in social media innovation: A microscopic analysis of Instagram advertising & its demographic reflection for buying activity with R. In *1st International scientific conference "modern management trends and the digital economy: From regional development to global economic growth"* (MTDE 2019). Atlantis Press.

18. Sharma, E., & Das, S. (2020). Measuring impact of Indian ports on environment and effectiveness of remedial measures towards environmental pollution. *International Journal of Environment and Waste Management, 25*(3), 356–380. doi: 10.1504/IJEWM.2019.10021787

19. Singh, I., Nayyar, A., Le, D.H., & Das, S. (2019). A conceptual analysis of internet banking users' perceptions. An Indian perceptive. *Revista ESPACIOS, 40*(14), 1–17.

20. Mohanty, P.C., Dash, M., Dash, M., & Das, S. (2019). A study on factors influencing training effectiveness. *Revista Espacios, 40*, 7–15. Retrieved from http://www.revistaespacios.com/a19v40n02/19400207.html

21. Singh, I., Nayyar, A., & Das, S. (2019). A study of antecedents of customer loyalty in banking & insurance sector and their impact on business performance. *Revista ESPACIOS, 40*(6), 11–28.

22. Gupta, D.K., Jena, D., Samantaray, A.K., & Das, S. (2019). HRD climate in selected public sector banks in India. *Revista ESPACIOS, 40*(11), 14–20.

23. Singh, S., & Das, S. (2018). Impact of post-merger and acquisition activities on the financial performance of banks: A study of Indian private sector and public sector banks. *Revista Espacios Magazine, 39*(26), 25.

24. Jain, S., Jain V., & Das, S. (2018). Relationship analysis between emotional intelligence and service quality with special evidences from Indian banking sector. *Revista ESPACIOS, 39*(33), 3–16.

25. Jain, A., Sharma, R., Dixit, G., & Tomar, V. (2013, April). Page ranking algorithms in web mining, limitations of existing methods and a new method for indexing web pages. In *2013 international conference on communication systems and network technologies* (pp. 640–645). IEEE.

26. Rehman, K.U., & Khan, M.N.A. (2013). The foremost guidelines for achieving higher ranking in search results through search engine optimization. *International Journal of Advanced Science and Technology, 52*(3), 101–110.

27. Cui, M., & Hu, S. (2011, September). Search engine optimization research for website promotion. In *2011 international conference of information technology, computer engineering and management sciences* (Vol. 4, pp. 100–103). IEEE.

28. Batra, N., Kumar, A., Singh, D., & Rajotia, R.N. (2014, February). Content based hidden web ranking algorithm (CHWRA). In *2014 IEEE international advance computing conference (IACC)* (pp. 586–589). IEEE.

29. Pardakhe, N.V., & Keole, R.R. (2013). Analysis of various web page ranking algorithms in web structure mining. *International Journal of Advanced Research in Computer and Communication Engineering, 2*(12), 4798–4802.

30. Chahal, P., Singh, M., & Kumar, S. (2013, March). Ranking of web documents using semantic similarity. In *2013 international conference on information systems and computer networks* (pp. 145–150). IEEE.

31. Nadanyiova, M., & Das, S. (2020). Millennials as a target segment of socially responsible communication within the business strategy. *Littera Scripta, 13*(1), 119–134. doi: 10.36708/Littera_Scripta2020/1/8

4

Link Building and Setup

4.1 Building, Reciprocal Linking, and Anchor Text

Links are the most important variables for all search engines for ranking and setting of web pages [1]. If you need to rank your website effectively, you will need to assemble links through link building. A good anchor text for SEO must have the following characteristics:

1. Succinct
2. Relevant to the linked-to page
3. Low keyword density (not overly keyword-heavy)
4. Not generic

4.1.1 Link Building for SEO

This is the method of getting new in-bound backlinks to a site from other websites. The main aim of third-party referencing efforts is to build future flow of traffic from Google, in spite of the fact that links from prominent destinations (for example, web articles) can likewise be a superb method of referral traffic. If one can generate more backlinks, then the chance of getting a high rank will be greater. Generally, each link is considered to be a vote in deciding the rank for a website [2]. So, more links means votes are higher, which ultimately helps in generating a higher rank. Anyway, it's anything but a "one person one vote" framework [3]. There are various variables that calculate the weight Google will give for a link and how it will impact future ranks, including:

1. **Control and Authority for Linking**

    When a website links to highly popular powerful websites, then it may give you a greater position which is considered as casting a ballot. So, a link from a highly ranked website (like ABC or CNN) will commonly be more effective than a link from a low-power site as it can give backlinks more effectively [4]. Here some tools are discussed which can control this link building.

a. *Ahrefs toolset* (a well-known toolset for backlinks and SEO analysis) has two metrics which determine the control and authority: *Domain rating (DR)* refers to the overall strength (authority) of a site and is measured from 0 to 100 – with 0 being no authority (for example a new site with no inbound links) and 100 being the highest – and *URL rating (UR)* refers to the strength (authority) of an individual page and is measured from 0 to 100 [5].

b. *Nofollow links (links from blog comments) may have the* **rel="nofollow"** *attribute applied as per Google's guidelines.* The nofollow quality tells web crawlers not to pursue the link and in principle doesn't add to rankings. Assessment changes anyway depending on whether this is really the situation. For instance, links from Wikipedia are nofollow and many driving SEOs consider Wikipedia links to be useful for rankings because of the site's high trust [6]. Third-party referencing master Eric Ward's sentiment on whether nofollow links add to rankings:

> I believe that any search engine with the goal of returning accurate results must maintain ultimate control over what it does and does not give credit to, and selectively choose the signals it values. And this includes rel=nofollow. I think it is perfectly reasonable to assume that any search engine can give credit to any link it wants to give credit to, if it feels there are enough signals to indicate it's a credible link. And that means even if the link is nofollowed. In my entire career I have never made a decision about whether I would pursue a link based upon whether that link would be followed or nofollowed. I pursue links if I feel they will help my user reach a certain goal or objective." While "dofollow" links are preferred, a high quality nofollow link may contribute to rankings and will certainly be a good link to acquire. Additionally, a mixture of "nofollow" and "dofollow" links will result in a more natural backlink profile.

c. *Side note*: While there is no "dofollow" link, SEO users use this term mostly for links when nofollow is not applied.

d. *Link placement*: The facts confirm that the web crawlers give more weight to specific sorts of links. For instance, a book link from inside content may confer more authority than a link from a sidebar or footer. This is on the grounds that an in-content link is bound to be either an editorial or clicked content or may be a link surrounded by appropriate text. Google's "Reasonable Surfer" patent, filed in 2004 and approved in 2010, proposes that Google will endeavor to decide how likely a link is to be "clicked" and select how much weight a link can have. That being stated, there is documented proof that Google keeps on offering weight to links from footers, sidebars, and so forth when the linking site is viewed [7].

Google uses external anchor text to help understand what your page is about and also, for which keywords it should rank. How do we know this?

Here's an excerpt from the original paper on which the Google algorithm is based:

FIGURE 4.1
Anchor text. Source: Author's screenshot

4.1.2 The Anchor Text

The search engines have long used anchor text – that is, the words used to link – as a ranking factor. This is the clickable link in the text which is hyperlinked and embedded in the content. Here in the screenshot example shown in Figure 4.1, the blue hyperlinked text is the anchor text. As per a study, over 16,000 keywords have been affirmed for good solid links between keywords and rankings [8].

So, the use of keywords in content, to have a strong rank in search, linking videos play a significant role. Otherwise keywords can't help in establishing links which will spoil the content and its role in getting a good rank. Too many content links can be an indication of control and may provoke search engine restrictions, for example, Google Penguin. It isn't significant content, but the link generation that matters. Yet, Google can make a differentiation between what it considers to be a typical appropriation of suitable content [9]. There are various types of anchor texts such as:

1. **Exact-match**: Anchor text is an "exact-match" in the search when it incorporates a keyword that represents the page that is being linked to. For instance: "third-party referencing" linking to a page with an appropriate external link.
2. **Partial-match**: Anchor text that incorporates a variety of the keywords on the linking page. For instance: "third-party referencing systems" linking to a page with an external link.
3. **Branded**: A brand name used as anchor text. For example: "Forbes" linking to an article on the Forbes Blog.
4. **Simple link**: A URL that is used as an anchor; "www.Forbes.com" is a simple link anchor.
5. **Generic**: A generic word or phrase that is used as the anchor. "Click here" is a common generic anchor.
6. **Images**: Whenever an image is linked, Google will use the text contained in the image's alt attribute as the anchor text.

4.1.3 Link Relevancy

There are various components that may impact the "importance" of a link and the weight that Google puts on it, including *(a) the anchor text and*

FIGURE 4.2
Anchor text with link relevancy. Source: Recreation from author's screenshots.

(b) other outgoing links on the webpage. It shows logically that the text surrounding the main text immediately gives a hint for ranking to search. Here the blue-colored links linked with the anchor text are showing link relevancy for the website. These links are also patented by Google under "Positioning dependent on reference settings" filed in 2004 [10] (Figure 4.2).

4.1.4 Unnatural Links

These links are flagged as unnatural by Google either by algorithm or manual supervision. These links don't contribute to ranking. Various unnatural links decrease the rankings/traffic. There is a touch of blurring of the lines in what builds up an unnatural link, with some argument that any link that is "built" is relied upon to control and manipulate data and this is unnatural. Too much manipulation and duplication will lead to misinterpretation by Google algorithm and a line routinely repeated by black hat techniques of SEOs to mistakenly legitimize their approach to manage external links. Any links were there has been an editorial option (the site has associated with you) are "white hat" and Google-safe [11]. This applies whether the linking site found your content organically (automatically by themselves), or you guided them toward it (through paid promotion/progression). Links gathered physically for traffic reasons (for example visitor blogging, content repurposing, content syndication) ought to comparably be verified when utilized as a piece of a continuously wide linking strategy and not exploited [12]. Basic link-building skills are used in good efficient link building in the post-Penguin age, combining different capacities, and the new age website must have a sponsor with:

1. An intelligent, descriptive point of view (contender assessment and framework)
2. An imaginative character (interface catch and content musings)
3. Excellent social capacities (frameworks organization and expanded linking)

It could even be said that convincing outer link foundation consolidates a segment of exploring the thinking element of the brain (working out how to get someone to interface with you). Before we continue forward to singular outside link foundation philosophies, we ought to explore two of the most noteworthy capacities in outsider referencing: Competitor research and expanded linking [13].

4.1.5 Competitor Research

A decent beginning stage for most third-party referencing efforts is to dissect the backlink profiles of contending sites who are in competition. This information can be accessed in our "alluding spaces" report (below), which demonstrates all areas linking to an individual website or site page by *Site Explorer > Enter domain or URL > Explore > Backlink profile > Referring domains*. The "Referring domains" report in Ahrefs will demonstrate to all of you the spaces that are linking to your objective site or URL [14]. Examining the backlink profiles of a few contenders will

1. Provide an insight into external link establishment methodologies
2. Give the opportunity to duplicate their links or links

Here the trick is to determine how or why the competitor got a link and decide if that is something you can duplicate. You'll most likely need to begin with some "quick successes". Here are a few instances of simpler links to reproduce:

1. *User-generated links*: Social profiles, video sites, blog comments, etc.
2. *Guest post links*: You can simply reach out to the linking site and ask if you can contribute your own guest post.
3. *A link from a resource page*: Reach out and ask if the site will add your resource.

Then search for "hubs" which will link to more than one competitor site. They can be very strong prospects. Harder to replicate will be genuine editorial links to popular content pieces on your competitor's site. You can use the "Best by Links" report in Ahrefs Site Explorer to quickly find the most linked-to content on a domain by *Site Explorer > Enter domain > Explore > Pages > Best by links*. To imitate these links, you will most likely need to "one up" your rival's content and that may lead to better ranking and linking. This may appear to be a ton of work, yet, depending on guidelines and procedures, developer can emphatically set the content to impact your rankings [15].

4.1.6 Technology of Outreach

This is ostensibly the most significant ability in a link developer's toolset, but at the same time is one of the hardest to ace. Except for uncommon cases – where content normally grabs viral footing and spreads independently of anyone else – to secure links that you should use in the content for better search and view in the best possible way. What's more, regardless of the ascent of social media, the best mechanism for outreach is still the e-mail.

Timing helps in the success of outreach as reacting quickly to generate and close pertinent queries is a fantastic technique to grow your outreach "hit rate". Hit rate helps in generating clicks that has liking interface with old published or most circulated articles. Mostly hit rate set up will be done by the second option whose steps are as follows:

1. Links to specific competitor content
2. Mentions your brand or a keyword you are tracking

These are great link-building opportunities, and early outreach can yield excellent results. Here if some developer links to the competitor's content, then reach out to them and request a link for your website too, or if someone publishes an article about a future product of yours, then request the writer to add your resources to the article. From visitor blogging to index entries, there are numerous methods for third-party referencing which will be talked about further in the book [16]. These fundamental focuses with regards to external link establishment must be considered to have importance for third-party referencing.

4.1.7 Reciprocal Linking

When it comes time to build links, you're likely going to be faced with a predicament known as reciprocal linking. At whatever point you attempt to procure a backlink on a site, regardless of whether it be through contributing a visitor article or something else, there is regularly a chance that the webmaster you're talking with is going to need to get a backlink from your site in return [17]. Sure, they get your extraordinary content, however, all things considered, you may be enticed to do this when somebody approaches you for a link on your site. This will at that point put you in a place to settle on a choice about equal linking. If you don't permit an included bit of content and backlink on your site, that site probably won't give you the opportunity on their site. As it were, you lose the organization and the backlink, if you don't respond. However, this link exchange or trade is what is beneficial for all parties. This is the central issue, and it's really fairly complicated too.

1. *When to Use Reciprocal Links for Success*

   In the past, reciprocal linking was all the more broadly utilized and acknowledged on the grounds that external link establishment was viewed as a decent practice. Today, it's critical to concentrate on visitors and not SEO. Finally, reciprocal linking truly can work for your perceivability just as for SEO purposes, so it's not something you ought to quickly reject. Still, there are a couple of various things that you should remember if you need it to work:

a. Only link when it is normal and important: The link which has no corresponding links. You should place an equally reciprocal link in the sidebar with the goal that it doesn't hold as much weight; however, this won't look normal to Google. The equivalent can be said in case you're linking to something that isn't important to your industry or site. This process has to be used in a perfect way so that Google will not penalize you.

b. Do not manufacture your campaign around this framework. It is anything but a smart idea to construct your whole external link establishment campaign around corresponding linking in light of the fact that at that point you're counterbalancing the majority of your diligent work (pretty much). Things are going to move quicker if you're building links and not giving them ideal backlinks. All things considered, the objective is to acquire links normally in the duplicate of the content while never requesting the same from back links..

c. It's typically not justified, despite any potential benefits, to respond to a contender. It's ideal if you can participate in complementary linking with an organization that complements yours and doesn't compete. For instance, in case you're a dentist, consider corresponding linking with a toothbrush company instead of another dentist. This will help ensure that the links are not going to rival one another.

d. Consider the authority of the other website. It's constantly a smart idea to check the authority of the site you may trade a link with. If your site has a PageRank 6 and a PageRank 3 needs a link on your site, it probably won't be justified, despite all the trouble to give a link to that page. The facts confirm that PageRank isn't all that matters, so there will absolutely be special cases, however this is a decent spot to begin. You would then be able to assess the site regarding quality and potential. In the event that the website isn't satisfactory, you would get a more fragile link than your rival would get, so be cautious.

It's critical to comprehend that, when you monitor backlinks, your links will be devalued if you're taking part in a great deal of equal linking. By the by, despite everything you're gaining in perceivability, regardless you are allowing an important group of spectators to tap on your link.

2. *When Reciprocal Links Are Best Left Untouched*

You must be extremely cautious with complementary linking in light of the fact that a lot of it can get you penalized by Google. Search engines normally see excessive linking as a black hat SEO strategy. Before, sites were trading links at an exceptionally quick pace, which was not helping visitors by any means. Google obviously put a stop to this and named it *"link farming"*. To make a long story short, it is OK if you need to take part in corresponding linking observing a portion of the guidelines above; however, don't do this constantly. A couple of complementary links to a great extent ought to be good enough to get you that link you needed on that extraordinary site, yet not put you on basic link formation [17].

4.2 Directing Traffic with Robot.txt Files

It is also known as the robot's rejection convention or standard. It is a book record that tells web robots (regularly search engines) which pages on the webpage can be crawled [18].

Search engines find robot.txt files as:

User agent: *
Disallow: /

The asterisk mark after "user agent" implies that the robots.txt record applies to all web robots that visit the webpage. The cut after "Disallow" advises the robot to not visit any pages on the site. For SERP, this is the best hack for SEO. If a web crawler crawls your webpage, it will search each and every one of your pages. Searching by Googlebot also helps in this process by two actions: By setting the limits of crawling and by which methods the crawling demand can be met. Essentially, the search spending plan is "the number of URLs Googlebot can and needs to crawl".

If you make the privileged robots.txt page, you can tell web search tool Googlebot to maintain a strategic distance from specific pages. In the event that you advise web index bots to just crawl your most valuable content, the bots will search and list your webpage dependent on that content alone. *No developer wants the server to be overcrowded by Google's crawler* [17,18].

4.2.1 Finding the Robots.txt File

In the event that you simply need to take a brisk look at the robots.txt file, there's a very simple approach available. One should type the URL of the site into the program's address bar like (www.xyz.com) and add /robots.txt to the end [19]. Then out of these three conditions, one will happen: (a) one will find the robots.txt file, or (b) one will find an empty file, or (c) one will get a 404 error.

If on second view you locate an unfilled document or a 404, you'll need to fix that. If one finds a legitimate record, it's most likely set to default settings that were made when one made the website [20]. When one finds the robots. txt file than it can be edited or you can delete all of the text but save the file. If one uses WordPress, then one can see this file when it goes to URL like yoursite.com as WordPress creates a virtual robots.txt file [21].

4.2.2 Creation of Robots.txt File

One can make another file by using a plain word processor only and erasing all previous history of the robots.txt file [22]. To begin with, you'll have to get comfortable with a portion of the language structure utilized in the creation. Google has a decent clarification on these terms [23].

First, a direct robots.txt file may be created and then modified. So, after starting with setting the user operator key, then one can apply this key to all by writing the syntax *"User-agent: *"*. Then write *"Disallow:"* here, as no restriction is there, website will be fine for all [24]. It looks very direct now. Then one can link the XML sitemap by typing *"Sitemap:* https://yousite.com/ sitemap.xml". The above XML sitemap resembles a fundamental robots.txt document [25].

4.3 Optimization of Robots.txt for SEO

For SEO, there are a wide range of approaches to utilizing robots.txt. One of the best uses of the robots.txt record is to augment search engine's expenditure plan by instructing them to not crawl the pieces of the website that aren't shown to the general population. *Type User-agent: * then Disallow: / wp-admin/ and finally write Allow: /wp-admin/admin-ajax.php.* It will help in creating backlink of the site; it wouldn't bode well for the crawler. So, to keep the crawler restricted, type *Disallow: /page/.* You may think in detail what sorts of pages to restrict from indexation [26]. Here are a few basic situations where that would occur: Purposeful copy content. If there is similar content in various places, then you can follow this. Also, one can use a *Thank You page to get into another lead. For this you need to apply a restriction after thank you by typing Disallow: /thank-you/* [27]. To make this all-inclusive, one can type *Noindex: /thank-you/* after Disallow. So, this page won't appear in SERP [28]. Be that as it may, the nofollow order is as yet educating web robots, so it's a similar idea. The important point is to find the exact place where the robots .txt can be placed [29]. Then the developer can find the source code of the particular page which the user wants to change and make sure it is in between the *<head>* <meta name=" robots" content=" nofollow"> <head>. There is no line in between. If both noindex and nofollow have to be given, then use *<meta name=" robots" content=" noindex, nofollow">.*

4.4 Content Writing and SEO-Friendly Content Writing

Content developing is well directed for adverts, keyword density, and keyword planning. All these help in website visibility and better optimization. This ultimately results in SERP and improves indexing along with ranking. The ideation and composing of content are the most important aspects of all [30]. By using headlines and sub-headers, adding links of previous content, optimizing the length, choosing the right keywords, optimizing images, and making content sharable, one can produce good qualitative content.

4.5 Link Title for Navigation

The link title attribute can be utilized to give more details to visitors who can click on content of their choice to browse [31]. Probably the noteworthy point on the web is that browsers don't have a clear idea where they are going when they follow links. Web browsers can display a short clarification of a link before the user chooses it. Such clarifications can give users a look at where the link will lead and improve their route:

1. Bad links are less inclined to be pursued; users will burn through less time.
2. Increasing users' comprehension of good links causes them to translate the web page where objectives are on display and decrease confusion.

In case you lay your cursor on the "information scent" link, the title will spring up after about a second. Having an enlightening tooltip show up when the mouse clicks on a link gives users a sign of the sort of data they can hope to get from following the link. Despite the fact that it is anything but a decent practice to depend just on the title credit to pass on information link, you can exploit it to give extra insight concerning that interface. The title attributes can be applied to other HTML components (for example, pictures and structure fields) close the links [32].

4.5.1 Guidelines for Link Titles

The objective of the link title is to enable users to anticipate what will occur if they pursue a link [33].

1. Appropriate data to incorporate into a link title can be:
   a. Name of the site the link will prompt (if not the same as the present site).

b. Name of the subsite the link will prompt (if remaining inside the present site yet moving to an alternate part of the site).

c. Added insights concerning the sort of data to be found on the goal page and how it identifies with the link name and with the setting of the present page.

d. Warnings about potential issues at the opposite end of the link (for instance, "Membership required" when linking to the Wall Street Journal).

2. Link titles ought to be under 80 characters, and should just once in a while go over 60 characters. Shorter link titles are better; anyway, super-tooltips are valuable for showing more content with regards to a huge picture display [33].

3. Do not add link titles to all links: If it is clear from the link name and its encompassing setting where the link will lead, at that point a tooltip will expand and at last decrease ease of use. (Regardless of whether users entirely hover on the link, the tooltip might be shown while the user moves the mouse.) A link title is pointless if it essentially makes use of # tag again, the same content or title that now appears in the link name shown on the page [34].

4. Do not expect that the link title will appear to be identical for all users. To be sure, screen readers won't show the content, yet may peruse the content so anyone might hear. Various programs will show link titles in various ways, as shown in Figure 4.2. What's more, most touchscreen programs won't show the link title by any means [35].

5. Finally, and most significant, link titles don't dispense with the requirement for good data: The link mark and its encompassing content ought to be reasonable regardless of whether the link title is shown. Users should not need to point to a link to comprehend what it implies: The tooltip can be held for strengthening data. Also, the link content must be appropriately organized to guarantee great searchability [36].

4.5.2 Touchscreen and Title Attribute

A few people recommend against link titles in light of the fact that crossing over to different display isn't upheld by touchscreen programs and by console-only programs. Although a unique signal, for example, the 3D Touch, could be utilized to show the tooltip, most touchscreen programs don't bolster this element. For whatever length of time that the link title is treated as an upgrade for mouse users and isn't required for utilizing the site, you can help a few users without harming others. Since a program that doesn't show link titles will basically disregard them and won't change in any capacity the format of the page, it is prescribed that you utilize this element as an approach to completely exploit the work area abilities and to develop linking for this device.

4.6 Google Webmaster Tools

Site developers everywhere throughout the world as of now use the Google Webmaster Tools to analyze, screen, and improve their webpages. Furthermore, the individuals who aren't utilizing this free element to further their potential benefit should begin right away. The knowledge you get from observing keywords, links, alarms, and manual activities is precious [30]. Here are five significant uses for Google Webmaster Tools:

1. *The Best Not-Provided Alternative for Keyword Data*

   Since analytic reports get into problems due to faulty messages, it is difficult to make sense of which keywords are the most beneficial. To begin, Google gives browser engagement and experience in terms of impressions and clicks and enables you to perceive how guests are arriving at your site. You additionally have information with respect to the active clicking factor, so you'll comprehend what percentage of individuals chose to tap on your site's link, and lastly the normal situation of your site in the list items [31]. Observing the keywords after some time may provide you some insight on what works for your site and what is the course you should take with respect to the keywords you have to rank on. The Search Queries tab related to the Google Analytics information will give you increasingly applicable data about your positioning keywords. Remember, the information given by Webmaster Tools isn't 100% complete, and they show just a little segment of keywords for destinations with heaps of positioning keywords [21].

2. *Monitor Your Site's Link Profile*

   The significance of the "Links to Your Site" included by Google Webmaster Tools can't be overestimated. You can see precisely what sites link to you (although this has a maximum breaking point, as well). For sites with in excess of 100,000 links, Google may not list all the backlinks. It's a blessing sent from heaven for SEOs in light of the fact that you can have hundreds of links with the information given in this device. You likewise have the choice to download the URL records in a CSV or a Google Docs document and use it in different backlink investigation forms [26].

3. *Get E-Mail Notifications about Your Site's Issues/Penalties*

   Get a split-second alarm you the minute you've been penalized by Google or if there are some other issues with your site. This is one of the bread-and-butter components since this is the main spot where you will see if there have been manual moves made against your site, in the event that you've been the objective of a pernicious programmer, or if there are robots.txt or crawling problems [28].

4. *Recover Your Site Using the Disavow Tool*

   It's not at all fun to work with Webmaster Tools' wrong settings if they don't have any kind of reasons. You may be hit with the penalties by the transcendent element called Google. In the event that your site is raising a great deal of warnings to Google, it will be investigated by a human and it will be penalized until the rule infringement is settled. In any case, don't let that spoil your otherwise good mood – it's not the apocalypse. Since Google Webmaster Tools (GWT) doesn't offer explicit data in regards to the links that were clicked on when the manual moves were made, you can browse through the links and see which one are unnatural and remove them [29].

5. *Identify Broken Pages on Your Site*

   As a marketing and promotion rule, if you attempt to get guests to your store yet you don't have the item on the rack, they are going to leave and make a beeline for your rival. It's similar with regards to websites. At the point when individuals can't get to your site or certain pages from your site, they'll return to the indexed lists page and click on another link. It renders all your SEO endeavors futile and no one needs that. Taking a look at the Crawl Errors included by Webmaster Tools now and again can spare you from disillusioning numerous guests. You'll have information about URL/webpage-related blunders that were recognized when Googlebot was crawling your site so you can fix the issues and improve the user experience.

4.7 Conclusion

By setting up your robots.txt document the correct way, you're not simply upgrading your very own SEO. You're additionally assisting your guests. In the event that search engine bots use the fund and plan very intelligently, they'll compose and show content in the SERPs in the most ideal manner, which means you'll be likely getting clicks. It additionally doesn't involve a lot of difficulties to set up your robots.txt document. It is generally a one-time arrangement, and you can roll out little developments as required. Building a good website that keeps on developing and performing well in search lists isn't simple. It requires a lot of focus on a lot of things; however, these devices make it significantly simpler. Website scrutiny helps in estimating different measurements of the website on different search engines. It empowers the developer to deal with the website in a superior way.

References

1. Baxter, L.R., Baxter, R., & Christen, P. (2003). *A comparison of fast blocking methods for record*. Washington, DC: KDD 2003 Workshops.
2. Hayama, T., Nanba, H., & Kunifuji, S. (2008, December). Structure extraction from presentation slide information. In *Pacific rim international conference on artificial intelligence* (pp. 678–687). Berlin, Heidelberg: Springer.
3. Fellegi, I.P., & Sunter, A.B. (1969). A theory for record linkage. *Journal of the American Statistical Links, 64*(328), 1183–1210.
4. Chakrabarti, K., Ganti, V., Han, J., & Xin, D. (2006, June). Ranking objects based on relationships. In *Proceedings of the 15th ACM international conference on information and knowledge management (CIKM'06)* (pp. 858–859).
5. Järvelin, K., & Kekäläinen, J. (2017, August). IR evaluation methods for retrieving highly relevant documents. In *ACM SIGIR Forum* (Vol. 51, No. 2, pp. 243–250). New York: ACM.
6. Giles, C.L., Bollacker, K.D., & Lawrence, S. (1998, May). CiteSeer: An automatic citation indexing system. In *Proceedings of the 3rd ACM conference on digital libraries* (pp. 89–98). Pittsburgh, PA: ACM.
7. Parsons, L., Haque, E., & Liu, H. (2004). Subspace clustering for high dimensional data: A review. *Acm Sigkdd Explorations Newsletter, 6*(1), 90–105.
8. McCallum, A., Nigam, K., & Ungar, L.H. (2000, August). Efficient clustering of high-dimensional data sets with application to reference matching. In *Proceedings of the sixth ACM SIGKDD international conference on knowledge discovery and data mining* (pp. 169–178), Boston.
9. McBryan, O.A. (1994, May). GENVL and WWWW: Tools for taming the web. In *Proceedings of the first international world wide web conference*, O. Nierstrasz (ed.) (Vol. 341). Geneva: CERN.
10. Chakrabarti, S., Dom, B., Raghavan, P., Rajagopalan, S., Gibson, D., & Kleinberg, J. (1998). Automatic resource compilation by analyzing hyperlink structure and associated text. *Computer Networks and ISDN Systems, 30*(1–7), 65–74.
11. Lawrence, S., Giles, C.L., & Bollacker, K. (1999). Digital libraries and autonomous citation indexing. *Computer, 32*(6), 67–71.
12. Shi, S., Xing, F., Zhu, M., Nie, Z., & Wen, J.R. (2006, November). Pseudo-anchor text extraction for searching vertical objects. In *Proceedings of the 15th ACM international conference on information and knowledge management* (pp. 371–382).
13. Shi, S., Song, R., & Wen, J.R. (2009). Latent additivity: Combining homogeneous evidence. *Proceedings of the 2009 workshop on text and citation analysis for scholarly digital libraries (ACL-IJCNLP'09)* (pp. 10–18), Suntec, Singapore, Tech. rep. Microsoft Research.
14. Robertson, S.E., Walker, S., Beaulieu, M., & Willett, P. (1999). Okapi at TREC-7: Automatic ad hoc, filtering, VLC and interactive track. *Nist Special Publication SP, 500*, 253–264.
15. Haveliwala, T.H., Gionis, A., Klein, D., & Indyk, P. (2002, May). Evaluating strategies for similarity search on the web. In *Proceedings of the 11th international conference on world wide web (WWW'02)* (pp. 432–442).
16. Nie, Z., Zhang, Y., Wen, J.R., & Ma, W.Y. (2005, May). Object-level ranking: bringing order to web objects. In *Proceedings of the 14th international conference on world wide web (WWW'05)* (pp. 567–574), Chiba, Japan.

17. Das, S., Nayyar, A., & Singh, I. (2019). An assessment of forerunners for customer loyalty in the selected financial sector by SEM approach toward their effect on business. *Data Technologies and Applications, 53*(4), 546–561.

18. Subhankar, D., & Anand, N. (2019, May). Digital sustainability in social media innovation: A microscopic analysis of Instagram advertising & its demographic reflection for buying activity with R. In *1st International scientific conference "modern management trends and the digital economy: From regional development to global economic growth" (MTDE 2019)*. Atlantis Press.

19. Singh, I., Nayyar, A., Le, D.H., & Das, S. (2019). A conceptual analysis of internet banking users' perceptions. An Indian perceptive. *Revista ESPACIOS, 40*(14), 1–17.

20. Mohanty, P.C., Dash, M., Dash, M., & Das, S. (2019). A study on factors influencing training effectiveness. *Revista Espacios, 40,* 7–15. Retrieved from http://www.revistaespacios.com/a19v40n02/19400207.html

21. Singh, I., Nayyar, A., & Das, S. (2019). A study of antecedents of customer loyalty in banking & insurance sector and their impact on business performance. *Revista ESPACIOS, 40*(06), 11–28.

22. Gupta, D.K., Jena, D., Samantaray, A.K., & Das, S. (2019). HRD climate in selected public sector banks in India. *Revista ESPACIOS, 40*(11), 14–20.

23. Singh, S., Mondal, S., Singh, L.B., Sahoo, K.K., & Das, S. (2020). An empirical evidence study of consumer perception and socioeconomic profiles for digital stores in Vietnam. *Sustainability, 12*(5), 1716.

24. Singh, L.B., Mondal, S.R., & Das, S. (2020). Human resource practices & their observed significance for Indian SMEs. *Revista ESPACIOS, 41*(07). Retrieved from http://www.revistaespacios.com/a20v41n07/20410715.html

25. Sharma, E., & Das, S. (2020). Measuring impact of Indian ports on environment and effectiveness of remedial measures towards environmental pollution. *International Journal of Environment and Waste Management, 25*(3), 356–380. doi: 10.1504/IJEWM.2019.10021787

26. Das, S. (2020). Innovations in digital banking service brand equity and millennial consumerism. In Kamaljeet Sandhu (ed.), *Digital transformation and innovative services for business and learning* (pp. 62–79). Pennsylvania, PA: IGI Global.

27. Mondal, S.R. (2020). A systematic study for digital innovation in management education: An integrated approach towards problem-based learning in Vietnam. In Kamaljeet Sandhu (ed.), *Digital innovations for customer engagement, management, and organizational improvement* (pp. 104–120). Pennsylvania, PA: IGI Global.

28. Singh, S., & Das, S. (2018). Impact of post-merger and acquisition activities on the financial performance of banks: a study of Indian private sector and public sector banks. *Revista Espacios Magazine, 39*(26), 25.

29. McCallum, A., Nigam, K., Rennie, J., & Seymore, K. (1999, July). A machine learning approach to building domain-specific search engines. In *Proceedings of the 16th international joint conference on Artificial intelligence (IJCAI'99)* (Vol. 99, pp. 662–667).

30. Baroni, M., Bernardini, S., Ferraresi, A., & Zanchetta, E. (2009). The WaCky wide web: A collection of very large linguistically processed web-crawled corpora. *Language resources and evaluation, 43*(3), 209–226.

31. Irick, K., DeBole, M., Narayanan, V., & Gayasen, A. (2008, April). A hardware efficient support vector machine architecture for FPGA. In *2008 16th international symposium on field programmable custom computing machines* (pp. 304–305). IEEE.

32. Davison, B.D. (2000, July). Topical locality in the web. In *Proceedings of the 23rd annual international ACM SIGIR conference on research and development in information retrieval* (pp. 272–279).

33. Amitay, E. (1998, August). Using common hypertext links to identify the best phrasal description of target web documents. In *Proceedings of the ACM SIGIR 98 workshop on hypertext IR for the web (SIGIR'98)* (Vol. 98), Melbourne.

34. Attardi, G., Gullì, A., & Sebastiani, F. (1999, May). Theseus: Categorization by context. In *Proceedings of the 8th international world wide web conference* (pp. 136–137). New York: ACM Press.

35. Nanba, H., & Okumura, M. (1999, July). Towards multi-paper summarization using reference information. In *IJCAI* (Vol. 99, pp. 926–931).

36. Nadanyiova, M., & Das, S. (2020). Millennials as a target segment of socially responsible communication within the business strategy. *Littera Scripta, 13*(1), 119–134. doi: 10.36708/Littera_Scripta2020/1/8

5

Concept of Off-Page SEO

5.1 Off-Page SEO

This deals with those activities where external links with different partners are established for improvement of search engine results page (SERP). These techniques deal with more complexity and specially with past web composition where listing or ranking can be done with establishing backlinks. Web crawlers use on-page SEO for depicting the information. Off-page techniques act on the principle of how the external world looks to the website from other links through web-based networking media (Facebook likes, tweets, Pins, and so forth.). It uses bookmarking to be shared among networks. It is useful in social media web-based sharing and generating significant traffic. A productive off-page SEO approach will act easily in favorable circumstances for webpage owners by

1. Increasing the ranking: Site will have a high rank in the SERPs and get more traffic.
2. Increasing page ranking: Page rank denotes the importance on a scale of 0– 10 in Google. Off-page SEO enhances this. It is one of 250 elements used by Google for ranking the sites. A higher ranking gives more significant clicks and help in achieving good rank for the website. Online networking will be higher too. It acts as a seamless action for webpage enhancement.

5.2 Backlink

This is one of the most used cult words in SEO development by worldwide developers [1]. Backlinks simply connect site pages. Blogging is nowadays used as an enhancer of backlinks. Previously blogging was used to position the web pages as with lot of backlinks which propel the ranking and positioning across all search engines. Here are some terms associated with backlinks:

1. **Linking juice or interface juice**: At the time of connecting two sites or blogs with the landing page, some juice or information is passed. This information is used to position the article and improve space authorization.

2. **Link of nofollow**: When a website links to another website, but the link has a nofollow tag, that link does not pass link juice. Nofollow links are not useful concerning the ranking of a page as they do not contribute anything. In general, a webmaster uses the nofollow tag when he/she is linking out to an unreliable site.

3. **Link of follow-up**: Those links, at the time of connecting two websites or blogs, which are important for future linking are called follow-up links.

4. **Root domain links**: Backlinks which come from an exceptional area even if the website is connected or not to multiple sites. It is the one and only connection of root space.

5. **Links of low quality**: These are links from automated websites, digitally harvested sites, spam, and even pornography sites. These play havoc if followed in the off-page SEO.

6. **Links of internal connections**: Those links which move within a small space internally are called inward links used for interlinking.

7. **Anchor text**: Hyperlinked text in the content as a grappled backlink which helps in the ranking of keywords.

5.3 Importance of Backlinks in SEO

Backlinks always improve the positioning as per Google's Penguin calculations. Logical qualitative backlinks from legitimate applicable websites or blogs always enhance the position. The reasons of importance are given below:

1. **They improve organic ranking**: They improve the ranking of the crawler. If content is getting natural links, then the website will be higher ranked. So, the objective will always be to get higher ranks, and individual backlinks along with those leading to the homepage.

2. **Indexing at faster rate**: Connection and slithering help backlinks to speed up the indexing as they have the exclusive connecting information.

3. **Traffic from reference**: They help in generating references for traffic inflow. How quickly references reply determines the speed.

5.4 How to Start Getting Backlinks

The nature and quality of backlinks determine their success. So paid connections may not be so helpful. So, quality backlinks can be got from:

1. **Writing good quality articles**: This is the most beneficial, least difficult approach for getting backlinks for a blog. Great quality organic content gets individuals connecting to it normally. Instructional exercises and top-ten rundown articles are several genuine instances of the kinds of posts that have incredible potential for getting referral backlinks [2].

2. **Start commenting**: Comments are a phenomenal, simple, and free approach to get backlinks. Start commenting on dofollow links in subsequent websites and pages. The most recent news has recommended that nofollow connections don't make much of a difference, yet start commenting on any blog, and you will definitely profit by some great linking juice. Commenting allows you to get strong single-direction backlinks as well as more traffic and better web search tool perceivability.

3. **Submit to web indexes**: This is nowadays not so followed due to the unavailability of a legalized web directory. So, in the absence of it, automated websites couldn't identify the backlink and this link will become a spam.

5.5 Dofollow vs. Nofollow Backlinks

Actually nofollow connections stop the connection limit (its quality and worth) from spreading to different pages. Then again dofollow connections disclose to Googlebots (at times alluded to as bugs) to follow the connection and list the page, in this way giving the connected page esteem.

Nofollow link: If a link is set to nofollow, it means the Googlebot won't crawl the link and won't pass any link juice. This is how nofollow links look:

<a href= http://www.example1.com *rel="nofollow">Example1</a>*

<a href= http://www.example2.com *rel="external nofollow">Example2</a>*

rel HTML tag: In both connections one has a keyword for nofollow in their genuine tag, which means they are the exact links for nofollow. Despite the fact that a nofollow connection doesn't pass any linking juice, it causes you to bring traffic towards your site. So, there isn't much negative activity in making a link for nofollow [3]. Google doesn't follow these connections; however other web crawlers regard them. You can likewise make a whole post nofollow with the assistance of a robots meta tag. For that you simply need to add the accompanying proclamation to the header area of a page/post. The programming is shown as *<meta name="robots" content="nofollow" />*. By adding the above statement a developer can make a whole post/page nofollow. All the links present on the page will be treated as nofollow even though they won't have a nofollow rel tag.

Dofollow link: A dofollow link passes link juice and helps in achieving a good positioning and ranking effect that instructs search engine spiders to follow itself. The programming is represented as *<a href* = http://www.example3.org *rel="external">Example3</a>*. By and large, when you post comments on different websites, you will see that the connections with rel outer quality, rel="external", are simply to speak to that the connection is outside, it has nothing to do with backlink type. If you discover no *rel quality* in connection even, at that point it is being treated as a dofollow link.

5.5.1 How to Check Whether a Link Is Dofollow or Nofollow

There are numerous approaches to check dofollow and nofollow links. The least demanding path is to right click on the connection and then click on "review component" and it will demonstrate to you the HTML code for your backlink. You can, without much of a stretch, check *rel property* and decide the connection type.

5.5.2 How Do Search Engines Interpret Dofollow and Nofollow Links?

All search engines that utilize the nofollow are worth avoiding as links that use it from their positioning calculation. This results in the shifting of interpretation from web index to web search tool.

1. Google states that their search engine takes "nofollow" literally and doesn't follow the connection by any stretch of the imagination. In any case, tests led by individuals show clashing outcomes. These examinations uncover that Google follows the connection; however, it doesn't list the connected-to page, except if it is in Google's record as of now for different reasons (for example, other links of non-nofollow that point to the page).

2. Yahoo! follows it, but excludes it from their positioning estimation.

3. Bing respects "nofollow" and doesn't include the connection in their positioning; anyway it isn't shown if Bing follows the connection.

4. Ask.com likewise respects the *rel quality*.

5.6 Submissions

Getting recorded in Google and the other prominent web crawlers is one of the best methods for coordinating natural (or all the more precisely unpaid), directed traffic to your site. Natural traffic is as yet the most important traffic on the planet, in 2017, with web search tools still appraised as the most confided-in hotspot for discovering news and data. The most prominent web search tools are Google, with around 90% of the pie, Bing, and Yahoo!

5.6.1 Submission Process

Manual submission gives the crawler the information that new data is there to be submitted to search engines, since crawlers have simplified the manual submission process so that spam can be prohibited. Once the developer has complied with all the guidelines, the website can be submitted to different search engines. In Google, the developer can submit by typing the website URL in Google's site page and verifying that a human is uploading, not a robot, and click to include the URL. The full guide is also available in Webmaster Tools. In Yahoo!, we have to seek the assistance of Microsoft's Bing as it has been helping Yahoo! from 2010. So, when we submit to Bing, it also automatically goes to Yahoo! In Bing also, like Google, the submission is done. After login and with the Bing Webmaster Tools you can submit the URL of the website.

5.6.1.1 Connect Site with Google Search Console

Sooner or later, to rank better in web indexes, you will need to get different sites to connect, so you should consider that first connection on outer destinations. In 2019, that generally means making valuable, exact, and in-depth content i that pulls in connecting links normally. If there is a need to avoid all that, until further notice, you can present your site and check it in Google Webmaster Tools. The technique to interface your site is straightforward with a little specialized information.

5.7 Directory Submission in SEO

Registry accommodation can be viewed as a repository or fountainhead of building safe backlinks. Blog indexes are regularly ignored and thought to be nasty, yet whenever done securely these are extraordinary undiscovered storage of dofollow backlinks in addition to introductions for your web journals (Figure 5.1).

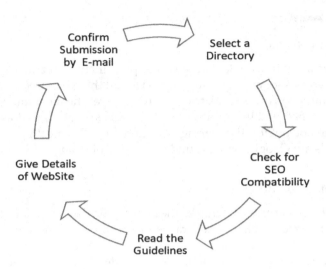

FIGURE 5.1
Directory submission process. Source: Author diagrams.

1. **Blog directories**: Wikipedia describes blog catalogs as huge data-bases which help sites dependent on classes and subcategories. All catalogs take the entire site as one blog. Paid blogs have their advantage in selection.

2. **Generating quality and safe backlinks**: Manual submission of blog entries doesn't involve spam and are a potential source of building SEO friendly backlinks for off-page SEO along with dofollow up links.

Here are the steps by which directory submission for SEO works:

Step 1: Pre-accommodation: This is a manual and direct method where a Notepad document will be prepared, then a blog URL and RSS channel and social media links and a summary of the website.

Step 2: Get some large setup for blog registries.

Step 3: Select the blog catalogs dependent on website decisions, similar to a lot of web journals' related PR activities where backlinks can be generated.

Step 4: Select a registry name.

Step 5: Select a significant space.

Step 6: Explore for best sub class on space.

Step 7: Select the sub class and classification that accommodate the blog's uniqueness.

Step 8: Click on the appropriate *Sub Category* and then click on *Submit Your Site* on the top right-hand side of the page.

Step 9: Fill in the details from notepad on the portal describing the URL, e-mail, CAPTCHA, and basic information.

Step 10: Submit.

Certain tips for safe approach

1. Don't go for ill-advised unsuitable links in place of posts and content.
2. Select and submit to the most suitable sub class.
3. Don't spam the index.
4. Don't give the wrong address or blog URL.
5. Don't prefer for quick backlink-generating repositories.

Self-submission: You should consider presenting your site to catalogs. There are some great-quality registries out there. A connection on *one or Two* of these indexes will presumably get your site into web indexes. Lamentably, there are a *lot* of low-quality links you will most likely unearth first. That sort of backlink can rapidly turn lethal – and a large number of them can cause you issues. Evade all minimal effort, modest SEO accommodation administrations, particularly those based around presenting your site to registries.

5.8 Article Submission for SEO

This action conveys a solid weighting in search engine optimization. It should be done with different assignments to be completely viable. Use the content cautiously for further subtleties in the most effective method to do article accommodation for SEO for greatest effect.

5.8.1 Reasons

1. Appropriate article accommodation work can be profoundly effective in boosting your rankings, yet the importance is in being "legitimate".
2. A connection from Page A of Website A to Page B of Website B conveys more weight if both Page A and Page B are about a similar subject. So, what better way to do the linking than write an article about something, similar like *"oak wood flooring"*, incorporate a link to your page about oak wood flooring, and submit it to as many article sites as will have it. At that point you would have heaps of pages on various sites, about oak wood deck, and all linking to your oak wood ground surface page.

There used to be a large number of sites that would acknowledge articles with practically no reviewing. At the most essential, the end article destinations would just request that the content you submitted to them was unique. That lead to text rewriting, composing what was basically a similar article about *"oak wood flooring"* yet in a marginally extraordinary way [4].

Web search tools understood this action was gaining power. Nobody at any point visited these article sites but to submit more articles to them in any case, and they were simple for web crawler calculations to spot – a large number of short articles, all with connections to only one webpage and ineffectively frequented by genuine browsers. If articles were spun by computers, then computers can detect the spinning so that the 50 so-called "unique" articles about "oak wood flooring" were anything but difficult to see as similar content. The inadequately controlled article sites were named *"content homesteads"* and any connections from them will be limited and useless [5].

5.8.2 Article Accommodation

Article accommodation sometimes appears as a shock to the SEO development but it is perfectly fine. Although numerous people who do SEO accept that article composing has been struck down, it was just one type of article composing that got hit – low quality. Article sites like Wikipedia and e-How were rarely impacted. With no or fewer examples these articles can be controlled better as pages connected to different sources. When they positioned in the query items web clients acted emphatically. These sorts of article sites proceed today and article accommodation is what they welcome. But at this point, you must have something genuinely to state so as to have the content acknowledged [6].

5.8.3 Article Destinations with "nofollow" Links

To demoralize spammers, many article sites mark their connections "no follow". They advise the web crawlers not to pass any interlinking juice from their website to the connected webpages. Anyway, this doesn't mean web crawlers absolutely overlook the connection, rather some just accept it as a factor. So, if you have got a link to your site from Wikipedia, Google gives you a pat on the back; however the pats on the head are not determined dependent on the authority of that Wikipedia page, they are doled out basically in light of the fact that you have the connection. This perception isn't generally acknowledged inside the SEO community or group [7]. However, it has been seen that it works over and over. It is perfect logic from a common-sense point of view that if you are developing a web index, wouldn't you consider a connection from a webpage like Wikipedia as a positive factor? Yes, perhaps the answer lies in the practice of logic.

5.8.4 Article Accommodation and Article Expansion

It is significant that probably the best article destinations enable you to alter their content or submit altering recommendations. Along these lines, for instance, you can alter a Wikipedia article with new or better data and connect this back to your page that contains the backup proof without presenting a whole article [8].

5.9 Press Release Writing and SEO Submissions

A press release is the data released to the media and public on some platform. Technically it can be described as "A Press Release or Press Release Submission is the accommodation of any new occasion or occurring of business over the public relation (PR) sites". An online press release is an official statement released to make the development public and can be accessed for free by anybody [9]. Advantages of an official press release are as follows:

1. *Gives authority locally and globally*: The admin who is releasing the information on a public platform will automatically get authority and power. The reliability and trust will rise and it gives immense authority in the industry.

2. *Gives distinctive visibility*: Press release sites give good perceivability and link profile to the admin for attracting huge traffic. So basically, make sure to share official articulations in regular intervals which are newsworthy and to optimize the first two paragraphs of a press release with appropriate and important keywords without flooding them. Visibility is due to the fact that the underlying two entries are essential for any official articulation.

3. *Gives a good relationship with media*: Statements and data in the public domain always give a bit of push to the relationship with managers of media. So official statements can be perceived properly, if you have a good relationship with a media house.

4. *Attracts investors*: Investors like to have transparency and trust in the system. So official statements, if presented and circulated properly with reliability, will automatically attract investors.

5. *Provides platform to increase sales and good relations with customers*: Transparency, trust, and clarity in a press release officially always give a positive image of the company to the customer, which boosts sales and get a good image in the customer's mind.

6. *Gives permanent index*: A regular practice of releasing information provides a permanent web index as visitors will search for an archive of releases and it will always be in the public domain.

5.10 SEO Plugin Installation

These days there are numerous stages accessible for blogging like WordPress, Ghost, Joomla, and some more. In any case, WordPress is the most well-known blogging stage since it is easy to use and gives numerous modules. We will consider the fundamental technique for introducing a module on the WordPress stage. So also, you can introduce different modules including SEO modules on the WordPress stage [10]. If you are utilizing WordPress.com, at that point you can't introduce modules. It is on the grounds that the users are utilizing WordPress.com, which has its constraints. To utilize modules, you should utilize the self-facilitated WordPress.org [11].

5.10.1 Instructions to Install a WordPress Plugin

There are three strategies for introducing a WordPress module: The first one is introducing a WordPress module utilizing search, the second one is transferring a WordPress module, and the third is physically introducing a WordPress module utilizing file transfer protocol (FTP).

5.10.2 Introduce a Plugin Utilizing WordPress Plugin Search

Module search is the least demanding. The main disadvantage is that the module must be in the WordPress registry which has limited free modules. So, *go to WordPress administrator and Click on Plugins >> Add New.* You will see a screen like the one in the screen capture in Figure 5.2. Find the plugin by typing the plugin name or the functionality you are looking for. After that, you will see a bunch of listings like the model shown in Figure 5.2.

You can pick the module that is best for you. Since, in the model, we were searching for Floating Social Bar which happens to be the first module, we will tap the "Introduce Now" button. WordPress will presently download and introduce the module for you. After this, you will see the completion message with a link to enact the module or come back to the module installer (Figure 5.2).

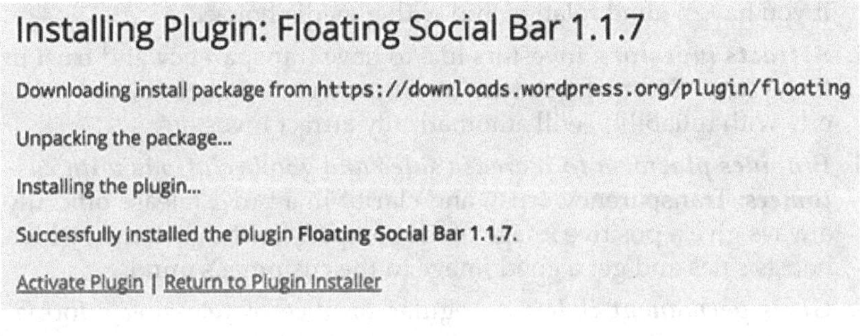

Installing Plugin: Floating Social Bar 1.1.7

Downloading install package from https://downloads.wordpress.org/plugin/floating

Unpacking the package...

Installing the plugin...

Successfully installed the plugin **Floating Social Bar 1.1.7.**

Activate Plugin | Return to Plugin Installer

FIGURE 5.2
Plugins installation. Source: Recreation from author login screenshots.

A WordPress module can be introduced on your site, yet it won't work unless you initiate it. So go ahead and click on the activate plugin link to activate the plugin on your WordPress site. That is all; you have effectively introduced your first WordPress module. The following stage is to arrange the module settings which will differ for each module [12].

5.10.3 Install a Plugin Using the WordPress Admin Plugin Upload

Paid WordPress modules are not recorded in the WordPress module registry. These modules can't be introduced utilizing the past technique. That is the reason WordPress has the upload technique to introduce such modules. So, to introduce WordPress modules utilizing the transfer alternative in the administrator territory we need to go very systematically. In the first place, you have to download the module from the source (which will be a compress record). Next, you have to go to WordPress administrator territory and visit *Plugins >> Add New Page, then click on Upload Plugin button over the page.* Then click on the choose file button and select the desired plugin file downloaded earlier. Then, once you have selected the plugin, tap on Introduce New Catch. WordPress will transfer the module from PC and display the success screen. Then tap on the Active Plugin connection to use the module [13].

5.10.4 Physical Installation of WordPress Plugin by FTP

When the facility provider for WordPress has limitations on the number of pages uploaded, then the admin will have difficulty in uploading the module. So here by using FTP only the plugin can be installed. First you will need to download the plugin's source file (it will be a zip file). Next, you need to extract the zip file on your computer. Extracting the plugin zip file will create a new folder with the same name. This is the folder that you need to manually upload to your website using an FTP client. If the username and password are not known, then by contacting the facility provider, one can get them and use them. Then, after entering FTP credentials, one can go by the path of *way/ wp-content/modules/organizer on your web server.* For transferring the modules, after visiting the WordPress administrator, click on plugins interface for the new addition. Once you click on *activate interface* under the module, then you can adjust the settings and the module can be activated [14].

5.11 Blog Posting and Comment Writing for Ideal Visibility and to Get More Followers

Visit the most mainstream writes blogs in your vertical and leave comments. As of late, comment blogging has fallen off in prevalence, in enormous part since comment spam prompted most websites to nofollow their connections. For off-page SEO, links are needed to a lesser extent. Your objective is basically to be the place the activity is and to leave an insightful comment that may grab the attention of the blogger, the digital platform's supervisor, and any influencers who may peruse the content. The equivalent applies for destinations like Reddit and Quora, where you can follow subjects explicit to your image or vertical and rapidly be seen as learned, astute, and compassionate in addressing others' inquiries or driving dialog. The SEO subreddit is exceptionally prominent, frequented by numerous individuals with huge names in SEO [15]. The relationships formed on these platforms have a way of paying huge dividends and can be invaluable for off-page SEO and reputation management. Regularly somebody sees your comments on one of these stages, begins tailing you there, and afterwards does a Google or LinkedIn search to find out about you or your image, which at last leads them to your site, where they may follow your newsletter or buy in to your blog.

5.11.1 Quit Guest Blogging for Links

Your image's success relies on a great deal of elements outside your site. Rather than visitor blogging exclusively for connections, utilize this strategy to enable you to construct an affinity with a portion of the top distributers, editors, influencers, and brands on the web. Whenever done effectively, the links do come. Be that as it may, as long as you make links clearly which all need, you're searching for a value-based relationship rather than one that is commonly useful, the harder it'll be for you to utilize visitor blogging successfully for off-page SEO. Peruse, leave comments on their websites, and interface with the top distributers in your vertical – or distributers that spread your vertical. When you have built up a compatibility and, ideally, have notoriety for making quality content, go for making a guest blog accessible for your operative platform. Regardless of whether that entryway opens, you'll have the option to sharpen your pitch and, in the end, get a foot in the door with different distributers. Keep in mind that your site is nevertheless one modest fish in a huge sea of alternatives. You have to interface with others in numerous spots off-webpage to construct the range and impact that will drive consideration and visits to your website, which is the place visitor blogging can be a gigantic resource [16] (Figure 5.3).

5.11.2 SEO and Guest Posting

As an essential target you shouldn't guest post. In any case, actually the roundabout effect becomes very powerful. There is nothing like building your reputation and perceivability to make individuals need a greater amount of your content. You get the opportunity to develop your very own audience, and at last a portion of these individuals will discover their way to your site, discover incredible content there, and link to it [17].

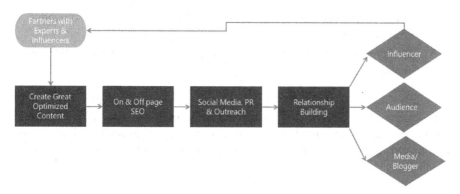

FIGURE 5.3
Content creation and relationship building. Source: Recreation from Author's conception.

5.12 Content

At the point when the vast majority state, think, or express "content", they frequently think content, pictures, recordings, and data shared by means of internet-based life. Truly, content speaks to the aggregate of the experience your brand designs, makes, and offers on the web and off, from logos and slogans to work force, telephone calls, signage, blog content, pictures, recordings, etc. If a prospect or client can connect with it, it is no doubt content [18]. There are many types of content present. With regards to off-site SEO, the primary concern you should concentrate on as to content is to see past content. You've no doubt heard that mobile like compact device is eating up the world. Indeed, think about what those untold millions are doing on their cell phones. To a great extent they are recording, which is expected to represent 85% of the content being shared online by 2019. Recordings and pictures can be a colossal boon to your image's off-page SEO, to a great extent since the two of them can be a low-venture/high-potential vehicle used to drive awareness and traffic back to your site, upgrading your off-page notoriety all the while [19]. Some minimal-effort, low-exertion approaches to utilize video and pictures include:

1. Post how-to recordings on YouTube, which is the second-most visited site on the planet and the #2 internet search engine behind parent organization Google. Also, mainstream recordings can and do rank in the SERPs.

2. Share day-by-day insights or recaps of distillations of recent events by means of Facebook Live.

3. Snap meeting or office funnies by means of Snapchat.

4. Compose picture collections on Imgur, Flickr, and other prominent photograph sharing sites.

5. Tweet recordings of fascinating things you see for the duration of the day.

Video and images are an incredible approach to demonstrate some character and make your image feel human, genuine, and alive to individuals who probably won't have known about your organization, or who've as of late found it. They can likewise do some amazing things for your off-site SEO [20].

For instance, say your brand has found that people who visit your articles have an improved probability of becoming real clients. You may shoot video meetings of the subjects or customers included in the most prominent sites, at that point present the recordings on YouTube, in addition to sharing them with the clients to post on their site, and spread by means of online networking. This improves the probability of considerably more individuals seeing the recordings and wanting to become familiar with your image [21].

5.13 Classifieds Posting: Classified SEO Ads: Do They Really Matter?

5.13.1 Classified Advertising

In straightforward terms, the advertisements that you see on papers, which can likewise be found on online magazines, news portals classified ads, etc. can enable you to support your items or blog's presentation at a lower cost; they are ordinarily much less expensive when contrasted with other showcasing methodologies. So, basically classified SEO ads are usually free to use and used to increase your products or blog's exposure.

5.13.2 How Classified Ads Are Helpful

For traffic mobilization and advertisement streaming, SEO developers use these classified adverts cautiously. It is revolving entity for traffic for generating benefits, likes, and clicks. Paid adverts always bring traffic, but that depends on how big the budget is; it is not for all the time. So, free adverts and organic placements of them are important. They don't cost anything and can start providing results from the moment they are employed if done with precision and if the content is substantial [22].

5.13.3 Three Sections of Free Adverts

1. **Headline:** An astute creative successful content-specific headline is what all adverts want as it attracts visitors and makes expert use of keywords that generate interest and bring in much desired readers and buyers.
2. **Details**: Subtitles give a detailed explanation of adverts for a product or service to different visitors of various language backgrounds.
3. **Detail advert subtleties**: Keywords like "Contact now" or "Buy now" give the final touch to the adverts.

5.13.4 Benefits of Classified Ads

Arranged advertisements are normally used to produce leads and increase click through rates. They are a phenomenal strategy for getting applicable back connections to your blog or site. A few sites enable you to post the content with back connections, and such ordered advertisements can quickly produce a quick turnaround in sales enquiry within hours of being posted. This is the reason utilizing free grouped advertisements to advance and market your online business is one of the most well-known and viable SEO techniques, utilized by most online advertisers [23].

Advertising your item and administrations through free online business advertisements is a capable method for connecting with your select group of potential clients. The thing to remember is to compose significant content that empowers the web crawlers to see unmistakably what your site is about, in light of the fact that that will enable them to rank you appropriately. This thus can give you better odds of being listed in the top few, with regards to search engines [24].

Characterized advertisements are loaded up with SEO potential as they utilize quite certain procedures. In the event that you are utilizing paid or free classified ads on your site, at that point your promotion and content get naturally filed on web indexes [25].

5.13.5 Five Benefits of Classified SEO Ads

Paid classified SEO advertisements are anything but difficult to utilize: Whether you know it or not, grouped SEO promotions are normally both versatile and search-benevolent. They are simple to explore and give your clients what they need. They are so brief but then give your group of spectators what they anticipate. This is the thing that makes arranged promotions use easy.

They open up a worldwide market for your items: Do you have any items or administrations to sell from your sites? Don't have the foggiest idea how to get more presentation for your items? Would you like to build your item deals? At that point, arranged publicizing is the best approach. It can make a worldwide market for your business and there's typically *huge* potential online to expand your item deals.

Most characterized SEO advertisements are free: There are a great deal of sites out there which help you post arranged promotions for *free*. Sites like OLX, Click.in, *Myadmonster*, and so on can help you effectively post free ordered SEO advertisements to support your items' perceivability. You don't need to contribute even a penny on promoting, so when you get everything for practically free, why not make the most out of them?

Make promotions in minutes not weeks: If you watch offline newspaper classified ads closely, most disconnected paper grouped advertisements set aside a great deal of effort to get distributed. For example, you need to address the paper promotion organizations to get your advertisement posted, and you additionally need to provide all the details. All these procedures for the most part take a great deal of time, and in addition these disconnected grouped promotions don't keep going for long, unlike online characterized SEO advertisements [26].

Search engine optimization for ordered sites can build your traffic: If you have an item-based selling site, you can colossally expand your traffic by posting your blog's or site address as characterized SEO advertisements, or then again you can essentially return a link to your destinations, so if somebody gets inquisitive about your items or administrations while perusing on the web, paid advertisements can arrive on your sites [27].

5.13.6 SEO for Classified Websites: What to Keep in Mind?

Here are some fundamental things to remember with arranged SEO advertisements.

1. Do not fill your site with paid advertisements that don't convey great content as they will do almost nothing for your business or SEO. All it will furnish you with is a high "skip" score, which won't be beneficial for you or your business [28].

2. A "bob" implies that individuals are going to your site yet are leaving without opening another page or tapping on the back connections that permit the site to capture the internet search engine's interest. It demonstrates that the content was bad enough to lead the client to continue with his question and his/her interest [29].

3. Make sure to utilize the sites that enable you to use back connections and direct click.

4. Do not post free arranged promotions each and every day. It is ideal to utilize them sparingly to produce the most extreme effect. This will likewise spare your site from getting the questionable title of *"ordered promotion site as it were"*.

5. Always make sure to post the data that you might want to promote explicitly, like expos, gatherings, social occasions, displays, and a wide range of open occasions as they will in general capture open public interest rather effectively.

6. The arranged promotions ought to be utilized shrewdly and ought to depict your organization and its administrations in the right level of detail. They should just contain the significant and applicable data. There is no space for blunders here. So, it is ideal to check, verify, and triple check the promotion before posting it on your site.

7. Much the same as some other procedure of boosting SEO rating, grouped SEO promotions also work just when utilized in the right way. Take great consideration and select promotions with back connections that will be effectively picked up by search engine tools and you won't be punished by any Google calculations.

8. Probably the least difficult approach to build your online presentation is to begin utilizing characterized SEO advertisements. Not only can you advance your business or sites on the web, you can likewise do so for *free*. You can likewise make others purchase your items or utilize characterized advertisements. Paid advertisements can absolutely improve your general SEO and give you an edge over your rivals. Be that as it may, ensure you are not doing it unreasonably, or else you may get punished by Google's most recent calculations [29].

5.14 Forum Posting

In the event that you are hoping to get great quality backlinks for your site, posting on forums is a trusted and simple way. You can utilize forum postings to expand traffic and improve the perceivability of your backlinks. Discussion posting is an intriguing off-page SEO procedure that causes you to build web traffic as well as increase your site backlinks too. Message sheets, dialog gatherings, talk sheets, discourse discussions, announcement sheets, and so on for the most part comprise as forum discussions. These forum postings act as a good platform to communicate with others in the same fields as well to increase your awareness in the growing market [30]. One can without much of a stretch look for direction from specialists on these gatherings. Countless great network gatherings exist on the web, and these can without much of a stretch assist you in learning and talking about numerous new things.

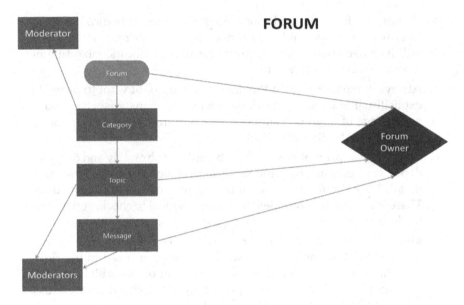

FIGURE 5.4
Forum posting. Source: Recreation from author's conception.

5.14.1 How Forums Work

Figure 5.4 shows the basic structure of a forum.

5.14.2 Dos and Don'ts of Forum Submissions

While posting on discussions you should be cautious with respect to specific things. Here are a few rules to enable you to utilize these discussions productively and helpfully:

1. **Be original**: Always post unique content. Posting distributed or replicated content is carefully restricted and can get you banned from the gathering.
2. **Focus on words**: Be cautious with the language. Never utilize hostile words or language. Compose with deference to every participant of the gathering.
3. **No direct promotion**: You can't utilize discussions as absolutely a showcasing instrument. Direct promoting in discussions is carefully precluded.
4. **Stay on topic**: Always adhere to your point to be credible.
5. **No spamming**: Spamming can get you suspended from the discussion, so carefully maintain a strategic distance from it.

5.14.3 Advantages of Posting on Forums

5.14.3.1 Create Connections

It empowers you to create associations with individuals who have comparable interests or who feel unequivocally about the issues that are near to your heart. By becoming a member of a forum, you can check what is happening in the discussion, what issues are being talked about, and what are the perspectives and assessments of different individuals. You can peruse the posts put together by individuals from the discussion. Additionally, you can present your very own posts and start an exchange.

5.14.3.2 Believability

Individuals accept other individuals' words more than direct, in-their-face publicizing [30]. So, by taking part in a gathering consistently and utilizing indirect marketing promotion methods you can fabricate believability for your thoughts, administration, or business.

5.14.3.3 Coaches

Discussion helps in discovering coaches that can offer you guidance on different issues and help you in various circumstances.

5.14.3.4 To Increase Website Traffic

Collections are an intriguing and simple approach to expand traffic to your site [31]. In the event that intriguing themes and subjects are talked about on your discussion, the individuals will keep coming back to check what's going on in the forum posts.

5.14.3.5 Building Connections

You can without much of a stretch form association with your guests with discussions. If you post routinely and effectively talk about various subjects with your discussion participants, then continuously, you'll become acquainted with them and all the more significantly they'll become more acquainted with you. Individuals are substantially more prone to purchase an item or administration when somebody they know sells or recommends it [31].

5.14.3.6 One-Way Link

Posting on collections give you an additional place to make a connection for your site in forum of signature links. Signature link is allowed by forums after the no. of posts after that forum member can signature link in his bio which appears at the bottom of every post [31].

5.14.4 Disadvantages

While forums are stacked with various advantages there are a few weaknesses connected to these discussions also. The principle detriment with discussions is the measure of time and exertion required to kick them off and look after them. It isn't simply constrained to getting individuals talking, you ought to direct it or moderate it also all through. This implies ensuring posts are fitting and that individuals are not spamming your gathering. This must be done every day, and if your discussion gets occupied it's will become very tedious. There are numerous gatherings everywhere throughout the web for you to make a commitment and add backlinks and learning. Simply pick your specialty and contribute, remembering the thoughts talked about above.

5.15 Business Listings

Professional resource sites for the most part have a huge amount of PR around their site and for the most part urge clients to drop a live connection alongside their profile. Having your site recorded on business listing destinations is a solid brand signal for Google as brand signs are turning into a bigger piece of the calculation. Give us a chance to utilize the prominent site Manta for an example of how to get backlinks from a professional resource site:

Step 1: First you have to sign up by filling in the required details.

Step 2: Then fill out the "Tell Us About Yourself" section of the site.

Step 3: Then you have to fill in the "tell us about your company" section.

Step 4: Scroll down below the paid options and click "No thanks. Continue to my profile".

Step 5: Fill in some basic details about yourself by clicking the edit option.

Step 6: Fill in some basic details about yourself. Near the website section click "edit".

Step 7: Lastly, enter the homepage URL under "company website".

Step 8: Enter your homepage URL under "company website" and click on Save.

5.16 Social Bookmarking Sites

The main concern is to use social bookmarking sites by third-party referencing.

1. To start with, there's still a point in utilizing social bookmarking for its unique reason – to store your bookmarks on the web – however some third-party referencing advantages are obvious if you do the entire thing right. These days, most social bookmarking sites boycott accounts that are bookmarking only one single site, or a little range of destinations. So, if you're simply going to submit links from your own site, inevitably, you will get restricted.

2. To utilize social bookmarking sites as third-party referencing methods in any capacity you need to construct a normal-looking profile containing connections to various sites.

3. This may seem like a great deal of work, however with a decent approach it isn't. You just need to utilize these destinations like you normally would. Each time you discover a fascinating site or a post just bookmark it, so when the opportunity arrives to bookmark your very own site, you'll have an all-around set-up profile waiting for you.

5.17 Estimation

The times when social bookmarking would propel your site to the highest point of web indexes are a distant memory. These days, you should just regard it as a supplemental method or treat it like a strategy to fabricate connections to your other online tools – like free web journals, web 2.0 destinations, etc. Along these lines it really works. Presently, best of all, the more sites like these you have the less you need to stress over your profile looking real. That is on the grounds that it naturally will be, because of the enormous assortment of sites bookmarked, in spite of the fact that they are generally yours. The reality about external link establishment is that no one strategy will be a silver bullet. If you do only article entries, or only web 2.0 sites, it won't get you far. What works is an assortment. Destinations that sit on top spots in Google are ones with huge backlinking profiles comprised of a wide range of sources of connections. Social bookmarking is a significant piece of this. It's common that a well-known site has a major number of bookmarks. In the event that it doesn't it just looks suspicious and sends conflicting signals. For instance, in the event that a webpage has 1,000 connections from article indexes, 10,000 bookmark links, 800 connections from online journals, 100 connections from YouTube, and so forth, then it looks trustworthy. Be that as it may, in the event that the site just has 1,000 article catalog links, at that point it appears to be suspicious. So, the estimation of social bookmarking lies in its capacity to make your backlinking profile total.

There are numerous social bookmarking sites on the web, so you can go as large or as little as you prefer. In any case, clearly, the more destinations you utilize the more work it will require. StumbleUpon is a decent spot to begin, as it's one of the greatest and most regarded social bookmarking destinations. At that point you can grow once you become acclimated to how the whole thing works. If you need to utilize more destinations one after another, manual entries probably won't be the best idea. There are some computerized and semi-mechanized administrations (some paid), for instance, BookmarkingDemon, IMAutomator.com, Social Poster, and SocialMarker. com. The more prominent SB sites you utilize, the more normal your backlinking profile looks. In any case, it's dependent upon you to adhere to a meaningful boundary somewhere and choose what number of websites is sufficient. Trying too hard isn't great either.

5.18 Social Networking

Utilizing social networking sites to get backlinks

Web-based life is seen by a great deal of organizations as basically "pleasant to have", and for certain organizations this is valid. It probably won't be the best medium on which to spend the biggest proportion of their computerized advertising spending plan, yet with regards to site improvement, a specific measure of critical inclusion will regularly correspond well with the site's rankings. In spite of the fact that the connections on the social stages themselves aren't weighted exceptionally on web crawlers, as a group they do serve as one sign of a site's relative power [31]. However when you utilize online life to get your message and your image out into the world, you can gain links from other people who use these sites, and those connections do make a difference fundamentally for SEO.

React to demands for your mastery

In the event that you are now touting mastery in your social profiles, individuals will connect and inquire as to whether you have considerations regarding a matter. Without requesting a connection, you ought to consistently do your best to react compassionately and be as valuable as could be expected under the circumstances. Commonly, individuals are composing an article when they connect [30].

Discover individuals making master gatherings and tweet at them; at that point add them to a rundown of master roundups

Beyond simply adding individuals to records, connecting with individuals who are making master collections is probably the most effortless approach to get a connection. Just telling them you're keen on being a part of a gathering could acquire you interface later on. Once more, advancing your specialty mastery will just make individuals progressively more apt to connect you and the network.

All your profile connections are low-hanging links that truly do make a difference! Likewise, utilize your depiction on Facebook for strong connections

Despite the fact that each tweet and Facebook connection post may not enable your SEO, whereas profile links truly do make a difference and can help drive some traffic also. Try not to think little of basically jumping on however many web-based life sites as could be expected under the circumstances and sharing the connection to your site. A significant number of these profile connections are dofollow.

LinkedIn's distributions highlighted on your profile enable you to share pertinent content you've worked hard on as well as create content on LinkedIn that will connect to your more profound asset

By taking the best and most well-considered content you compose and including it in your distributions highlight, you are getting links; however, you'll likewise collect some traffic in the event that you are dynamic on LinkedIn.

Have a procedure that includes Tumblr and Pinterest as they permit connects and are commonly well-regarded

Tumblr and Pinterest are two significant social media sites in light of the fact that their connections are high-worth and better regarded, somewhat, more than Facebook and Twitter links. You can even have an If This Then That (IFTTT) formula that re-posts your blog entries on Tumblr in the event that you need to make it increasingly mechanized. More than connections, Pinterest drives noteworthy traffic when used fittingly, so there's another motivation to concentrate on it in your content procedure by and large.

5.19 Conclusion

For effective SEO to be done and dusted, off-page SEO is proven to be marksman's arrow along with WordPress. It enhances the actions of on-page SEO and places them in perfect sync. It is always advisable, when you go for third-party referencing, that you don't go by short-cuts. The difficulty level reaches a high when you go for connections with high clicks which has become a little more troublesome with a flood of .com sites. Social bookmarking also has less effectiveness. In a time of emergency, one can use sites like reddit. com, stumbleupon.com, scoop.it, and delicious.com (to give some examples) to develop the content.

References

1. Ankalkoti, P. (2017). Survey on search engine optimization tools & techniques. *Imperial Journal of Interdisciplinary, 3*, 40–43.
2. Beel, J., Gipp, B., & Wilde, E. (2010). Academic search engine optimization (aseo) optimizing scholarly literature for google scholar & co. *Journal of Scholarly Publishing, 41*(2), 176–190.
3. Benczur, A.A., Csalogany, K., Sarlos, T., & Uher, M. (2005, May). Spamrank–fully automatic link spam detection work in progress. In *Proceedings of the first international workshop on adversarial information retrieval on the web* (pp. 1–14).
4. Brin, S., & Page, L. (1998). The anatomy of a large-scale hypertextual web search engine. *Computer Networks and ISDN Systems*. doi:10.1016/s0169-7552(98)00110-x
5. Conway, S., & Sligar, C. (2002). *Unlocking knowledge assets*. Seattle, WA: Microsoft Press.
6. Couzin, G., & Grappone, J. (2006). *Search engine optimization: An hour a day*. San Francisco, CA: Sybex; Chichester: John Wiley Distributor
7. Cui, M., & Hu, S. (2011, September). Search engine optimization research for website promotion. In *2011 international conference of information technology, computer engineering and management sciences* (Vol. 4, pp. 100–103). IEEE.
8. Davis, H. (2006). *Search engine optimization*. O'Reilly Media, Inc.
9. Subhankar, D., & Anand, N. (2019, May). Digital sustainability in social media innovation: A microscopic analysis of Instagram advertising & its demographic reflection for buying activity with R. In *1st International scientific conference "modern management trends and the digital economy: From regional development to global economic growth" (MTDE 2019)*. Atlantis Press.
10. Singh, I., Nayyar, A., Le, D.H., & Das, S. (2019). A conceptual analysis of internet banking users' perceptions. An Indian perceptive. *Revista ESPACIOS, 40*(14), 1–17.
11. Mohanty, P.C., Dash, M., Dash, M., & Das, S. (2019). A study on factors influencing training effectiveness. *Revista Espacios, 40*, 7–15.
12. Singh, I., Nayyar, A., & Das, S. (2019). A study of antecedents of customer loyalty in banking & insurance sector and their impact on business performance. *Revista ESPACIOS, 40*(06), 11–28.
13. Gupta, D.K., Jena, D., Samantaray, A.K., & Das, S. (2019). HRD climate in selected public sector banks in India. *Revista ESPACIOS, 40*(11), 14–20.
14. Singh, S., & Das, S. (2018). Impact of post-merger and acquisition activities on the financial performance of banks: a study of Indian private sector and public sector banks. *Revista Espacios Magazine, 39*(26), 25.
15. Jain, S., Jain V., & Das, S. (2018). Relationship analysis between emotional intelligence and service quality with special evidences from Indian banking sector. *Revista ESPACIOS, 39*(33), 3–16.

16. Das, S., Mondal, S.R., Sahoo, K.K., Nayyar, A., & Musunuru, K. (2018). Study on impact of socioeconomic make up of Facebook users on purchasing behavior. *Revista Espacios, 39*, 28–42.

17. Mondal, S., Das, S., Musunuru, K., & Dash, M. (2017). Study on the factors affecting customer purchase activity in retail stores by confirmatory factor analysis. *Revista Espacios, 38*, 30–55.

18. Mondal, S., Mall, M., Mishra, U.S., & Sahoo, K. (2017). Investigating the factors affecting customer purchase activity in retail stores. *Revista ESPACIOS, 38*(57), 22–44

19. Singh, S., Mondal, S., Singh, L.B., Sahoo, K.K., & Das, S. (2020). An empirical evidence study of consumer perception and socioeconomic profiles for digital stores in Vietnam. *Sustainability, 12*(5), 1716.

20. Singh, L.B., Mondal, S.R., & Das, S. (2020). Human resource practices & their observed significance for Indian SMEs. *Revista ESPACIOS, 41*(07). Retrieved from http://www.revistaespacios.com/a20v41n07/20410715.html

21. Sharma, E., & Das, S. (2020). Measuring impact of Indian ports on environment and effectiveness of remedial measures towards environmental pollution. *International Journal of Environment and Waste Management, 25*(3), 356–380. doi: 10.1504/IJEWM.2019.10021787

22. Das, S. (2020). Innovations in digital banking service brand equity and millennial consumerism. In Kamaljeet Sandhu (ed.), *Digital transformation and innovative services for business and learning* (pp. 62–79). Pennsylvania, PA: IGI Global.

23. Mondal, S.R. (2020). A systematic study for digital innovation in management education: An integrated approach towards problem-based learning in Vietnam. In Kamaljeet Sandhu (ed.), *Digital innovations for customer engagement, management, and organizational improvement* (pp. 104–120). Pennsylvania, PA: IGI Global.

24. Nadanyiova, M., & Das, S. (2020). Millennials as a target segment of socially responsible communication within the business strategy. *Littera Scripta, 13*(1), 119–134. doi: 10.36708/Littera_Scripta2020/1/8

25. Di, W., Tian, L., Yan, B., Liyuan, W., & Yanhui, L. (2010, July). Study on SEO monitoring system based on keywords & links. In *2010 3rd international conference on computer science and information technology* (Vol. 5, pp. 450–453). IEEE.

26. Jain, A., & Dave, M. (2013). The role of backlinks in search engine ranking. *International Journal of Advanced Research in Computer Science and Software Engineering, 3*(4), pp. 21–25.

27. Kaur, M., & Singh, H. (2015). Web health analyzer for search engine optimization. *International Journal of Engineering and Technical Research, 3*(8), pp. 30–32.

28. Kent, P. (2006). *Pay per click search engine marketing for dummies.* Seattle, WA: John Wiley & Sons.

29. Ma, Z. (2006). *Web-based intelligent e-learning systems.* San Francisco, CA: IGI Global.

30. Poojary, S., Narayan, T., & Temkar, R. Search engine optimization analyzing & interpreting the necessity of SEO for enterprises. *International Advanced Research Journal in Science, Engineering and Technology, 4*(7), pp. 138–144. doi: 10.17148/IARJSET.2017.4724.

31. Žilinčan, J. (2018). Improving information accuracy with SEO for online marketing services. In Natalia Kryvinska and Michal Gregus (eds.), *Agile information business* (pp. 217–253). Singapore: Springer.

6

Search Engine Algorithm and Search Engine Marketing

6.1 Internet Search Engine Algorithms

In basic terms, an algorithm is a math equation that accepts a query as information and returns an answer for the issue, ordinarily subsequent to assessing various potential arrangements. A web index calculation utilizes potential keywords as the info issue, and returns important query items as the arrangement, coordinating these keywords to the outcomes stored in its database. These specific keywords are controlled via a web crawler that breaks down site page content and specific keywords' importance dependent on a math recipe that will differ starting with one internet browser, then onto the next. A web crawler is a tool for giving the possible output of the query given by the visitor to the search engine [1]. It does two precise activities like output reciprocation and ranking of data or results as per their importance so that SEO will have maximum impact. The impact of a website directly depends on the web crawler and its setup. The indexing of webpages plays a very vital role in setting up the correct web crawler. Rankings, highlights, and run-down results of webpages have a complete effect on the detailed web result. Web index calculations are paramount and very clearly help the settings to be precise. Web browsing is the root of web crawlers. Web indexes, due to their mysterious acts for crawlers, are often called the "Secret Sauce". Besides basic elements of web index, webpages including both on- and off-page factors play very vital role in browsing.

6.1.1 Mechanism of Web Crawlers

They have two important works to execute, of which crawling is the main function and the building of an index is the next best one. Crawling is the technique of allowing tools like robotized crawlers or bugs to arrange huge amounts of interconnected information from the web. The crawlers discover the pages and extract codes to store information from databases to be shown later. So, web crawler organizations have data centers to control this setup and they hold huge amounts of data too. So, when any significant demand comes for them, crawlers decide and move to set up accordingly within seconds. They are compatible with content, webpage, keywords, etc. On the other hand, index building is the technique of sorting out the information gathered and stored in *file* containing 100 million gigabytes of memory. This helps in extracting information as and when required by the crawler depending on the search query.

6.2 Algorithm

When a lot of information is accumulated on website pages, there may be duplication and replication. So, to have particular information, we need to have some sort of rule or mechanism for deriving accurate output. Here the algorithm plays the vital role. Calculation of perplexing and protracted nature often plays a vital role in making the algorithm for the web crawler [2]. These calculations help in effective SEO setup which improves the ranking. In an attempt to enhance their capacity for returning those correct answers rapidly, search engines started refreshing their radically changing dynamic calculation in which they conveyed applicable outcomes to browsers. Due to these adjustments in the calculation, numerous sites were punished with lower rankings, while different sites encountered a flood in natural traffic and improved rankings along these search results [3]. Think about a vehicle made during the 1950s. It may have an extraordinary motor, yet it may likewise be a motor that needs things like fuel infusion or it may not be able utilize unleaded fuel. At the point when Google changed to Hummingbird, maybe it dropped the old motor out of the vehicle and put in another one. To help each search user find the correct information they're looking for as quickly as possible. Updates in particular further cement search engines' responsibility to their visitors.

6.3 Need for an Update

Above all else, how about we start by examining the Google calculation? It's massively muddled and keeps on getting increasingly convoluted as Google tries its best to give searchers the data that they need. At the point when web search tools were first made, early search advertisers had the option to effortlessly discover approaches to make the web crawler imagine that their customer's webpage was the one that should rank well. Now and again it was as straightforward as placing in some code on the site called a meta keywords tag. The meta keywords tag would tell web search tools what the page

TABLE 6.1

Difference between Past Search and Google Hummingbird Search

Past Search Techniques	Google Hummingbird Search
Keywords-based search	Question-based, conversational, semantic search that takes into account why, when, who, etc.
A similar quest yields the same outcome for various users	The same capture may yield various outcomes to every client dependent on current conditions.
Search systems don't utilize client data for searching	It utilizes client data like past ventures, land area, social offers, and so on.
Past search techniques cannot deal with complete inquiry strings	By utilizing Google's information chart, it can deal with long complex capture strings.
Gives visitors numerous potential alternatives to proceed with search	It begins with addressing the inquiry. This is called conversational looking.

Source: Author's own interpretation and explanation.

was about. Previously, the Google calculation would change inconsistently. If a site was sitting at #1 for a specific keyword, it was ensured to remain there until the following update which probably wouldn't occur for quite a long time or months. At that point, they would push out another update and things would change. They would remain as such until the following update occurred. Three of the greatest changes that have occurred over the most recent couple of years are the Panda calculation, the Penguin calculation, and Hummingbird.

6.4 Google Hummingbird

The Hummingbird is a completely new calculation. It approaches search engine queries in a brand new and intelligent way using new innovation joined with more seasoned highlights of the current calculations. It is named for the speed and exactness of the little winged creature. Its quality is the capacity to rapidly examine longer, increasingly complex questions and give the most accurate responses to the searcher with the least potential clicks; it has a progressively human approach to interface with its visitors and give a more straightforward answer, not at all like its past Panda and Penguin renditions. The potential contrasts between past inquiry systems and Google Hummingbird are delineated in Table 6.1 [4].

6.4.1 Scarcely Any Methods for Advancing for Hummingbird

6.4.1.1 Try Not to Neglect Site Speed

Hummingbird needs to convey quick, exact outcomes. Thus, one can accept that a quick site would rank higher than one that takes numerous seconds to stack. Help Google see better the content of your site.

6.4.1.2 Concentrate on Semantics

Rich snippets help Google better comprehend the content on your page. This is fundamental. There are rich snippets for everything – use them. Snippet is a result that Google shows to the browser for the query.

One of the objectives of semantic web capture is to remove insignificant list items. Rich bits = Google understanding content = trust.

6.4.1.3 Making Incredible Content

Hummingbird shifts toward conversational capture and away from specific keywords. Expound on what your visitors would find supportive, need to know, have inquiries concerning, and so forth. In addition to the fact that this is an extraordinary advertising apparatus for your business, it can even now help search traffic.

In addition, Panda and Penguin were folded over into this new motor too. Being the supplier of answers that individuals are scanning for is the perfect situation for your site/business.

6.4.1.4 Convey and Concentrate on Intent

This strategy ties in intimately with content, as you need to distinguish needs, issues, arrangements, and after that expound on them. In any case, site use measurements will turn into a considerably increasingly well-known metric since Google is eliminating specific keyword information in analytics. Rather than concentrating on the specific keywords that are carrying visitors to your site, take a gander at how those prominent inquiry pages are performing [5]. What do they convey? Are visitors exploring your site like you want/anticipate? Once more, you're enhancing for visitors and *not* the web crawlers. These advancements are not new, and anyway the new algorithm update is driving site proprietors to do what Google has been recommending from the beginning: Center your visitors and make novel, accommodating content. Search engine optimization was (now and again still is) thought of as a "handy solution" and it's advancing into a great deal more.

6.5 Google Panda, Google Penguin, and Google EMD Update

Panda utilizes a calculation named after the Google engineer, Biswanath Panda. In February 2011, the primary capture channel that was a piece of the Panda update was launched. It's fundamentally a content quality channel that was focused at low-quality and slight sites with little SEO control, to keep them from positioning highly in Google's top web index results pages (SERPs) [6]. The Panda calculation update changed the SEO world. It changed the content system, specific keyword research and targeting. It even changed how corresponding links work, since amazing important connections indicating a page eventually add to its worth with regards to SEO control. Google could now decide all the more precisely which sites were "malicious" and which sites would probably be considered valuable by guests. Prior to Panda, poor content could rank exceptionally or even overwhelm Google's top outcomes pages. Panda 1.0 was released to fight content farms.

Panda 2.0 update: This update, launched in April 2011, was focused at worldwide inquiry inquiries, and additionally affected 2% of US search inquiries. So, Google continued refreshing and expelling the imperfections as well as improving the search engine to find accurate solutions. In May 2014, Google launched Panda 4.0 and it had a noteworthy effect. One of the sufferers was eBay. It lost a critical level of top-ten rankings it had previously delighted in. When you construct a site today, you need to reliably compose and distribute inside and out content. This content must include esteem, be intriguing to the visitor, and take care of accessibility on cell phones. If you neglect these, then you can't draw in visitors and the change rate will be low [6].

6.6 Google Penguin

Google Penguin refers to a lot of algorithmic updates and data refreshes for the Google internet search engine that the organization occasionally starts to help improve the estimation of its quest inquiry results for visitors. The Google Penguin updates fundamentally look to prevent different kinds of web crawler spam (otherwise called spamdexing or Black Hat SEO) from being effectively compensated as higher-placed web crawler results [7]. Web crawler spam can incorporate exercises, for example, keyword stuffing, connect spamming, the utilization of undetectable content on site pages, and duplication of copyrighted content from high-positioning sites, and that's just the beginning.

6.6.1 For What Reason Are Links Significant?

If connections are established with a site, then there is a chance that the whole website can be connected and it becomes a recommendation. If an obscure website is connected then there may not be too much benefit. In any case, if you can get an enormous number of these little ballots cast, they truly can have any kind of effect. This is the reason, previously, SEOs would attempt to get the largest number of connections as they could from any conceivable source. Another thing that is important in the Google algorithms is anchor text. Grapple content or anchor text is the content that is underlined in a connection [8]. Along these lines, in a connection to an SEO blog, the content would be "Website optimization blog". If ABC.com gets various sites connecting to them utilizing the grapple content "Web optimization blog", that is an indication to Google that individuals scanning for "Web optimization blog" most likely need to see sites like ABC in their indexed lists.

6.7 Recovering from Google Panda, Penguin, and EMD Update

If your site rankings have dropped starting lately, there are a few different kinds of penalties that could be impacting your site. Google's two essential calculations that emphasize rankings are called Panda and Penguin. Panda is the calculation that chooses the quality and congruity of your site's content, as it relates to specific keywords. Regardless, your site is often considered as the spirit of your business; it's your virtual retail facade. Having it disappear from Google (which is by a wide margin the most unmistakable web record) is unthinkable for most organizations. In case you've been punished, trying to recuperate in the correct way, in view of long haul objectives, is the best practice.

6.7.1 Google Panda Penalty Recovery

Sites that get punished by Google's Panda calculation get hit for one of two reasons: Replication of content or duplicate content [9]. This implies that your website pages are not uncommonly enchanting, helpful, or are copies of various pages on your site page (or all the more terrible, various sites). This kind of punishment is the more direct one to recover from, as it is controlled inside and Google's updates happen all the more regularly.

6.7.2 Google Penguin Penalty Recovery

If your site was suddenly hit hard and your rankings dropped radically, odds are that it was related to Google's Penguin calculation [9]. This is the most inconvenient punishment to recover from and amazingly, much of the time the frequently most widely recognized. Low-quality SEO firms routinely add a website to private online diary frameworks, low-quality libraries, and sites with an ultimate objective to trap the web searcher. This used to function admirably, and even today, a couple of associations see brief outcomes – yet it's anything but an enduring arrangement. The Penguin calculation definitely compensates for wasting time, debases the sites, and punishes sites that use these techniques.

6.8 EMD Update

When website admins were scrambling to manage three back-to-back Penguin updates, at the same time as Panda updates, Google discharged the EMD update channel [9]. The EMD update, also called "the accurate match space" update, algorithmically targets sites that utilize a definite match area to rank for specific keywords in Google. So, this update basically evacuates the authority ordinarily connected with these areas and now and again applies a penalty to them. In the event that the specific keyword you needed to target was "Search engine optimization blog" and you wanted to target "SEOblog.com" and were positioning high for this keyword, Google may punish this area, especially for that specific keyword. Before 2012, if you purchased the EMD adaptation of a specific keyword, you were going to rank for it very commonly. What numerous website admins and subsidiaries were doing was purchasing up hundreds or thousands of these domains, putting little presentation pages on them, and guiding traffic to associated items or facilitating Google AdSense advertisements. Indeed, even before the official EMD update, there were signs that Google was beginning to punish them, or possibly cheapen them. Since 2012, numerous things have changed inside Google's calculation. There have been significant updates, and a few extra activities they've used throughout the years. Many smart SEOs even detected the patent Google discharged about definite match areas. This might be the reason why some EMDs are appearing back up in the SERPs.

6.9 Staying away from EMD Update

Unfortunately, there truly isn't a step-by-step manual for recuperating from the EMD update. You either have an EMD or you don't. There are a couple of tips that can decrease the likelihood of getting hit, or conceivably recuperating from it:

a. Don't make exact anchor texts for your site.
b. If you do have exact match anchor text, disavow them preemptively.
c. Don't form accurate match entryway pages.
d. Don't stuff specific keywords of your EMD into your titles and meta tags.
e. Don't stuff specific keywords into your content.

If hit by the EMD update, odds are you have been punished for the specific keyword that is contained in your space [10]. If that is the situation, and your entire realm relies upon that one keyword, it is recommended that you move to another domain. However, Google didn't authoritatively declare this on their blog as they have with numerous other algorithmic movements.

6.10 Real Ranking Signals

Google is perpetually improving their algorithm and SERP identification. Since 2016, it has been improving with the help of AdWords which bring the highest revenue [11]. Google officially also confirmed this advert introduction and SERP enhancement in 2017 by implementing the following actions:

1. **Page Loading Speed**

   The client expects the website to stack as quickly as possible. People use cell phones to place a query more often than searching on PC. So, page burden has to be increased and it needs to be taken care of [11]. Google positioning is very important for this. To improve client experience, the website needs to be optimized so that it can stack in under 3–4 seconds [12], because after more than 3–4 seconds individuals will not wait for it to respond. Increasingly more traffic is originating from cell phones nowadays; those utilizing mobiles to peruse are for the most part in a hurry and don't care for keeping an eye out for a page to stack. A review uncovered that 40% of customers will forsake a site that takes over 3 seconds to stack; along these lines, if your site doesn't stack rapidly enough, at that point you're going to pass up potential visitors as they head off to your rivals. Gary Illyes from Google expressed that the page speed of your portable pages will be a significant factor with regards to the versatile benevolent calculation. Google as of now has a few instruments to help test and improve your site's page stacking speed which is a solid sign of its significance to the web crawler. There's likewise a versatile ease of use report inside the Google Search Console. Further uplifting news is that the Google Accelerated Mobile Pages (AMP) venture enables distributers to empower their pages to stack faster on cell phones. It is additionally anticipated that AMP pages will be progressively significant for mobiles in 2019 also.

2. **Nearby**

   A year or so ago, nearby outputs became little bit conspicuous in Google as more visitors use web index frequently with an increase of 33% over previous practices [12]. So, it needs significant attention. Regardless of whether your business is worldwide or localized, visitors will look for the local nearby addresses. So, it's critical to have them isolated out and filed accurately as well. Guarantee that the

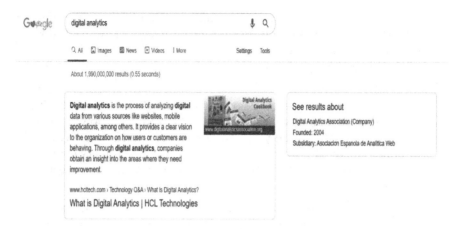

FIGURE 6.1
Digital analytics in www.google.com for snippet inclusion. Source: Recreation from author login screenshots.

data in your Google My Business listing(s) match the name, address, and phone number (NAP) on your site and in other nearby references around the web. This will invigorate your posting and will assist your site with ranking better for local searches.

3. **Snippet Inclusion**

Google currently shows highlighted links for search, and this has been gradually incorporated from 2017 with a high increase of adverts. For this setup one needs to appear above the organic results with a link to the source, so essentially it's better than having the number one result for that search query (Figure 6.1).

4. **Portability Index**

Different themed looks are always in demand since 2017. Google will soon be making their index mobile-first, meaning that desktop will no longer be their priority. Visitors need a versatile benevolent means of reciprocation to the queries. Zoom in and zoom out are also used. Advertisers always look for these things so that they can go ahead with the mobile bandwagon, best way of indexing and representing this insight since 2017. Versatile pursuits imply that there will be all the more of an emphasis on neighborhood look as well – something that organizations won't have any desire to pass up. Google has a helpful mobile portability test that can reveal to you whether your site is streamlined for versatility and even gives tips on the most effective method to improve it.

5. **Ranking Hampered by Pop-Ups**

From January 2017, Google started prohibiting pop-ups on portable websites. Up to August 2016, 85% of all webpages indexed in Google had the *versatile amicable mark*. Interstitials hamper the content as they can't draw traffic with their activity which makes the website slow [13]. Thus, to much appreciation of users, Google has responded to the utilization of these pop-ups by bringing down the position of results.

6. **Web Content**

Qualitative content began to be prioritized from 2011 by Google with the Panda calculation update. So low-quality content is downgraded and dropped. High-quality content draws good traffic and a higher ranking. The results are a case of Google advancing quality content. Google even made a *Webmaster Guidelines page* where they express the following:

Give excellent content on your pages, particularly your landing page. This is the absolutely most significant activity. If web pages contain helpful data, their content will pull in numerous guests and allure website admins to connection to your webpage. In making a supportive, data-rich site, compose pages that obviously and precisely depict your theme. Consider the words visitors would type to discover your pages and incorporate those words on your site.

Just as telling us what quality content is, Google unequivocally guide us to keep away from the accompanying content-related systems:

- Automatically created content
- Participating in connection plans
- Creating pages with almost no unique content
- Cloaking
- Sneaky sidetracks
- Hidden content or connections
- Doorway pages
- Scraped content
- Participating in associated projects without including adequate worth
- Loading pages with immaterial specific keywords
- Creating pages with malevolent conduct, for example, phishing or introducing infections, trojans, or other malware
- Abusing rich snippets mark-up
- Sending mechanized questions to Google

In their rules, Google clarify that they will make a manual move against your site if it doesn't agree to its rules.

7. **Connections**

These are important for crucial functions of websites from 2017. In 2016, Andrey Lipattsev, Search Quality Specialist for Google, stated that connections along with content are Google's top two ranking signals. Connecting links propel connections to various publishing adverts so that the content enhancement can also be done.

6.11 Search Engine Marketing Idea

Visitors generally browse by a registry chain called "chain mode" [14]. Web index along with advertisements frequently make the business. So, when someone is buying or giving feedback, the data to be filled is to be pertinent. Then it can be identified by web crawlers. This helps in making the traffic natural and more organic where unpaid SERPS can be done [12–14]. Generally, advertisers think of SERP as a paid outcome, but the reality is that listed items are what make the SERP authentic. Search engines along with trade sites with their potentiality in profitability make the engine a good source of online advertising. Along with various nations and their rules, search engines accommodate these regulations where it operates locally. So that the advertisement can reach the maximum audience in that country. Keywords play the all-important roles in identifying how a search query can better provide optimized output. So, to use the power of focused traffic, adverts look for how to use paid and natural search engine marketing (SEM) for getting get more outputs. Web index traffic is very much useful for unique promotional activities as it acts in different ways like:

1. **As a non-problematic technique for adverts**: Most advertisements, on- and offline, interact with traffic when the visitor visits the webpages and search engine taps the browsers towards the adverts. So how quickly and advert catches the attention of the person marks the success rate. It is very much an easy and not a problematic process of promotion.
2. **As an audience-driven voluntary drive for browsing and search**: Browsers browse with their set of keywords which has images, content, video, or other easily known things. So, search query helps us in getting more web index and results in generating more traffic for the content.
3. **As a natural PR tool**: Organic search engine marketing is based on the grounds that many web indexes base their significance calculations on content including image, video, etc., which explicitly act as a PR tool for future visitors to get maximum output.
4. **As a tool to improve web crawler**: Everything considered, the web crawler considers that there are still compelling reasons to put legitimate efforts behind organic SEO optimization, particularly in site design, content formatting, content clarity optimization, and server

platform adjustments. Paid adverts assume that they will be viewed by all the platforms equally. The accompanying paid adverts are generally normal and of different types:

a. Paid arrangement and incorporation

b. Shopping search

c. Video search promotions

d. Local search advertisements

e. Product posting advertisements

6.12 Pay Per Click and Quality Score

Advertisement by pay per click helps to bring more traffic as and when the client needs. It is a unique way to get more browsers and increase the traffic [15]. If not managed properly, one can draw a lot of traffic with heavy expenditure but with no profit as visitors will visit and leave; as a result, there is no conversion for the money spent. Pay per click (PPC) is very direct. Search engines help in generating paid ad postings or forum which show up nearby and are placed above non-paid natural queries from organic search. Search engines will be paid with every click on the listing. These promotion spots are sold in an auction (Figure 6.2).

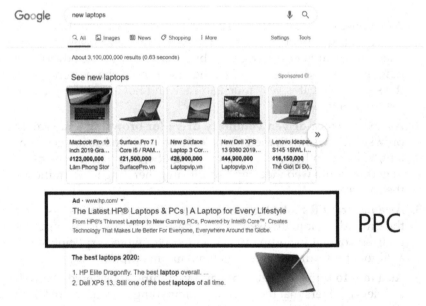

FIGURE 6.2
PPC and Quality Score for new laptop search. Source: Recreation from author screenshots.

You go for the maximum money that can be paid for a click on the ad. The maximum offer will propel the ad to be in the highest order. The positioning of the ad directly relates to the Quality Score. Clicking on a PPC post will lead to the page you have chosen. For keyword planner, if you offer $1.50 max, then that is what is called an elevated offer for max click. If 100 clicks come, then you will be paying $150.00 to the PPC administration. Due to its importance for promotion, PPC generates highly targeted traffic to your site for a fraction of the cost of any other form of paid advertising.

6.12.1 Importance of PPC

PPC can produce significant traffic instantly and quick promotion within a particular network with the help of AdWords. It is direct, and if paid well, then it will help the website to the top. So, links linked with ads will make the content visualized at the max priority. PPC advertising is also nimble whereas organic search engine marketing or other forms of advertising can lag weeks or months behind changing audience behavior, you can adjust most pay per click campaigns. PPC takes less time in enhancing the website. It can sometimes act as a deal to promote the website efficiently.

6.12.2 Challenges for PPC

The most challenging aspect of PPC is the keyword setup as that needs precision, foresight, and money to deliver efficient and effective adverts. Since keywords have a limited presence in the planner, it takes huge effort to set them up [16]. Unnecessary ambiguous traffic is another challenge for PPC, as PPC quickly increases the cost and limit the web speed for browsing due to faulty keyword in content. So, the content needs to be very much optimized with the speed and keyword for effective PPC to give better ROI. PPC has no scale of measurement [14–16]. For this reason, the client needs to pay heavily for the setup. Sometimes PPC may not be cost-effective for low-budget websites.

6.12.3 Advertisement and PPC

A lot of websites don't always depend on PPC for traffic generation as they know that sometimes the high cost can backfire [16]. So, they try to be smart and effective:

1. **Campaign and issue-oriented adverts**: For a short period of time if one website is in competition with another, then these situations can be tackled by PPC to create buzz in say 24–48 hours [17]. A limited time with a higher budget acts like a catalyst to propel the effect of PPC to a higher extent (Figure 6.3).

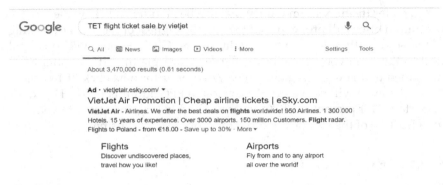

FIGURE 6.3
PPC and advertising. Source: Recreation from author screenshots.

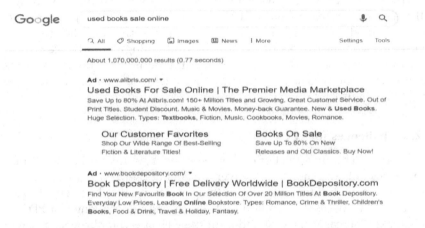

FIGURE 6.4
PPC and direct response business. Source: Recreation from author screenshots.

2. **Direct-reaction:** For the sale of an item or offer that people can buy immediately while searching in the website then, one should go for PPC. Online purchase and e-commerce are rampant in the use of this technique [17]. The ROI will be good, and it will be readily acceptable (Figure 6.4).

3. **B2B awareness:** For any business to business activities, the promotion can be done effectively within a business cycle with PPC [18]. It increases the perceivability and controls the duplicate adverts too (Figure 6.5).

4. **Niche market terms for product listing:** For extremely small explicit keywords, if you are trying to create traffic, then PPC is your one-stop solution [16–18]. For a particular brand of shoes with color and style, the browser search then will get all possible brands and prices at one place with their product listings. These adverts place the item

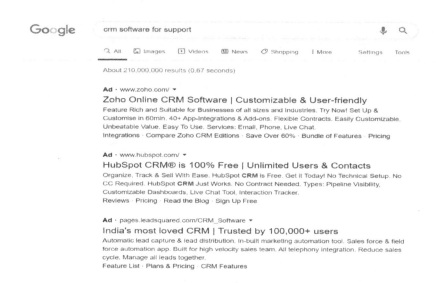

FIGURE 6.5
PPC and B2B awareness. Source: Recreation from author screenshots.

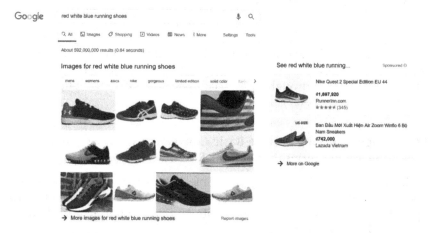

FIGURE 6.6
PPC and product listing. Source: Recreation from author login screenshots.

of a particular brand and provide the information for particular visitors under PPC (Figure 6.6).

5. **Remarketing**: Google AdWords plays a very vital role in this. It helps in creating an audience of users who have already visited your website [19]. Customization is very much possible with it. So, customers who once visit the website may return to the website if remarketed. If it is used with PPC then the cost may go down subsequently [20].

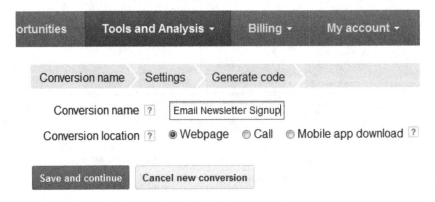

FIGURE 6.7
PPC and track conversions. Source: Recreation from author screenshots.

6. **Conversions matter, not the clicks**: By giving more consideration to transformations than to clicks, and tracking conversions where you need to remain on a spending plan, one needs to follow changes like:

   a. *Visitor makes a buy.*

   b. *Visitor finishes a business request structure.*

   c. *Visitor downloads a white paper and registers.*

A change needs to be significant. However, a transformation must mean something to you. If you can't think of a quantifiable, valuable result of a visit to the site, don't burn through cash on compensation per click publicizing – there's no point (Figure 6.7).

6.12.4 Google and Bing Provide Basic Conversion Tracking Within Their Ad Platforms, but Not for Revenue Not for Revenue

Google Analytics help in giving a tangible output to the PPC sources and CRM or client relationships like HubSpot. Here the nature of content which propels the visitor to turn to the website is given importance and subsequently the PPC goes to the revenue.

1. **Make a sensible realistic budget for PPC**: There is no rule of thumb for budget setting. It depends on how effectively you understand the conditions and objectives [20]. *Cost per click is not equal to change rate × total clicks × benefit per transformation. It is the money spent to get the benefit generated per click.*

2. **Policy of not using long-tail keywords**: Sometimes more extensive terms attract more browsing and therefore generate more search. In this regard, if long keywords are used then the promotion campaign

will get clicks but expenditure will be higher. So, it will not be a profitable method.

3. **Use explicit keywords**: Significant keywords always drive more traffic. So those keywords which have more extensive use will be more used.

4. **Never ignore good significant content**: Sometimes superlative words with more detailed explanation become eye-catching. These words must be used to attract more traffic. They not only get a good ranking but also drive effective traffic too.

5. **Quality score**: Search engines like Google and Bing have Quality Scores which look after advertisement, keywords, point of arrival duplicate, click measurements, on-location use measurements, and so on. They act on the history of website search, adverts duplication, put keywords in adverts and the landing page, split long keywords into smaller ones, and focus on the traffic for the website.

6.13 Cost Per Mille (CPM)

CPM stands for cost per mille (mille is the Latin word for "thousand", signifying "1,000 impressions"). CPM is identified with CPC, yet, as opposed to being charged when somebody clicks on your promotion or link, you are charged each time your advertisement just shows up on a page [21]. You would run a campaign of CPM if your objective was to bring issues to light or possibly test another medium or channel; however you should be cautious with them. They can go nonstop, demonstrate your advertisement all over, and you can rapidly end up running up an enormous cost for advertisement with little to show for it. In any case, as the world has turned out to be progressively mobile and portable, research studies have demonstrated that upwards of 60% of advertising clicks on mobile are accidental. That being the situation, CPM or CPA (see beneath) could demonstrate to be increasingly beneficial.

6.13.1 Cost Per Action (CPA)

You possibly pay when explicit activities happen, for example, a buy, a web structure is submitted, and so forth. CPA expects the "activity" is past the click. When running a campaign for CPA you would build up objectives that could be followed, and after that pay (likely a higher sum than CPC) when those objectives are accomplished.

6.13.2 Return on Investment (ROI)

When you put your well-deserved cash into anything – crude materials, staff, contractual workers, promoting – you do as such with the desire that those speculation dollars will return and bring along a great deal of "companions". Perhaps your most ideal method for developing your income is to contract more sales reps or pay higher commissions to joint endeavor accomplices that will advance for your benefit. The primary concern is that you should follow how you spend your cash if you are not kidding about developing your business, income, and productivity.

6.14 Change Rate

One of the most significant pieces of paid consideration campaigns, the transformation pace of a site is the estimation of the achievement of a paid incorporation venture. A transformation rate is estimated by the quantity of potential guests playing out the ideal activity, regardless of whether the activity is purchasing an item, rounding out a structure, or some other objective of the page. For instance, if there are 100 guests to a site page by means of compensation for every click on an advertisement, and one of those 100 purchases the item the site sells, at that point the change rate for that promotion is 1%. The bigger the transformation pace of a site page, the more fruitful the site will be, just as how effective the paid incorporation venture is. Estimating change rates can be extremely hard for organizations who are utilizing paid incorporation to carry more business to the website page. For the most part, pages use web investigation projects to quantify the transformation rates they are getting when they launch another compensation per click campaign. The web examination programming ponders the conduct of site guests, and gathers information for use in a wide range of research territories, including finding out about flow change rates. Information used to examine transformation rates is contrasted against key execution markers, so as to decide whether the transformation pace of the site is at the pinnacle rate or if there are changes that should be made to the venture to attain the ideal outcomes [22].

A significant piece of change rates is amplifying the odds of the guest turning into a transformation. This can occur in a few different ways. One of the approaches is to make the guest progressively defenseless to change and coordinate the correct guest with the perfect time and the correct site. This is hard to do; however that is the point at which such things as paid consideration and site improvement become possibly the most important factor. Website pages can likewise augment the odds of a guest becoming a conversion by making the ideal activity, regardless of whether it is rounding out a structure or purchasing an item, as simple as could reasonably be expected. Guests react to the simplicity of the activity, particularly if it is likewise advantageous to the guest. Offering an advantage to the guest can likewise raise the transformation pace of the site. There are numerous approaches to raise transformation rates, a large portion of which have to do with being everything a potential client needs [19,21]. Obviously, improving change rates may simply have something to do with perceiving the capability of the site. A portion of different approaches to boost the odds of a change include:

1. Knowing the one of a kind selling purpose of the site page.
2. Being memorable with unique information.
3. Offering a few installment alternatives.
4. Having a reasonable returns strategy.
5. Having clear approaches on data assurance.
6. Being clear and brief with data

A transformation rate of conversion can be one of the most significant factors in a paid campaign. The proportion at which the cash poured into a promotion effort is justified, despite all the trouble, or not is surely something that any business will need to know.

6.15 Strategizing PPC Campaigns and Understanding Performance Metrics

6.15.1 Dealing with Your PPC Campaigns

The benefit of paid adverts is that you can have your site recorded on the primary pages in a noticeable spot on Google and other web indexes. In any case, showing up is just a piece of the procedure. You have to make an advertisement that prompts clicks, but converts to sales or whatever outcome you're searching for. In the event that you don't have the foggiest idea what you're doing, it's conceivable to compose a promotion that individuals are attracted to and click on, however, you don't make sales. Since you pay per click, and clicks can add up rapidly, you can lose cash. When you've made your new venture or web visit, you'll have to oversee it consistently to ensure it keeps on being successful. You ought to upgrade your campaigns:

1. **Include PPC keywords**: Expand the reach of your PPC campaigns by adding keywords that are significant to your business. Use Wordtracker for keyword research, and you'll discover it's especially useful for discovering negative specific keywords. SEOs are

principally worried about keywords they can incorporate into their duplicate, yet in PPC you have distinctive match types – wide, expression, and exact. If this is your first PPC venture you can utilize expansive matches to attempt to perceive what the market is doing and what individuals are typing in. Utilize accurate match when you are certain for the best specific keywords which will give you the desired result. Another prevalent procedure is to utilize expansive and expression matches with a great deal of negative keywords.

2. **Negative keywords**: Non-changing over terms as negative specific keywords to improve venture importance and diminish squandered spend.

3. **Split ad groups**: Improve active visitors' click-through rate (CTR) and Quality Score by separating your promotion bunches into littler, progressively pertinent advertisements, which help you make more focused-on advertisement content and points of arrival. PPC campaigns enable you to have diverse advertisement varieties, so you can have content promotions which can build your CTR. By creating various types of promotion, you can screen any examples demonstrating which advertisement is being clicked, and which aren't [23].

4. **Survey costly PPC keywords**: If survey is costly and PPC keywords are failing to meet expectations, then specific keywords can be shut off.

5. **Refine landing pages**: Modify the content and invitations to make your greeting pages line up with individual pursuit inquiries to support transformation rates. Try not to send all your traffic to a similar page. A great deal goes into structuring a triumphant PPC venture: From investigating and choosing the correct specific keywords, to arranging those keywords into efficient campaigns and promotion gatherings, to setting up PPC points of arrival that are improved for changes. Web indexes compensate publicists who can make significant, intelligently targeted pay-per-click compensation per-click ventures by charging them less for promotion clicks. If your advertisements and greeting pages are valuable and fulfilling to visitors, Google charges you less per click, prompting higher benefits for your business. So, in the event that you need to begin utilizing PPC, it's essential to figure out how to do it right. Every one of your items ought to have its own page, with an extremely clear "catch" and a decent depiction. Amazon do this in great way. Link the promotion directly to that page – the keyword should reflect the advertisement, which should reflect the point of arrival. It just takes individuals 3 to 4 seconds to settle on their choice so you need to ensure the page rushes to stack [23]. Some significant hints while proceeding with PPC campaign are recorded below:

    a. Never bid on costly specific keywords regardless of whether they are well-known keywords.

    b. Check routinely to ensure that the majority of your advertisements are dynamic.

    c. Start your specific keyword offering at the base sum offered by the pay per click organization.

    d. Always remain inside your spending limit.

    e. Never pay for keywords that won't carry guests to your site.

    f. Monitor your campaign continually and make changes when required.

6. **Execution measurements**:

Some of the important metrics and their meanings are listed below:

Visits: This is a visit to your site. An individual may see only one or numerous pages when they visit.

Impressions: This is how frequently your advertisement is shown for a given specific keyword in the given timeframe.

Clicks: This is how frequently an advertisement was clicked. When all is said and done, this is compared with visits, yet they may not generally coordinate. For example, if your advertisement is shown, however, you additionally appear in the natural postings and somebody clicks that interface, this will bring about a visit without a click. For another situation, a searcher may as of now be on your site and play out another inquiry in a different tab or window. If they click a promotion there, it will bring about a click, yet not an extra visit.

Cost: This is the expense of a keyword or advertisement bunch for the given time.

Click through rate (CTR): This measurement is processed by taking the quantity of clicks and separating by the quantity of impressions. About 1.75% is the normal among all enterprises; however, this will rely upon the uniqueness of your keyword [24].

Cost per click (CPC): This is the normal expense per click for a keyword or a promotion gathering. It is determined by taking the total cost and isolating the quantity of clicks.

Revenue per click (RPC): This depends on a set-up revenue incentive to a change. In the event that no revenue value is assigned, at that point the RPC will be $0.00.

Return on investment (ROI): This is figured by isolating RPC by cost, and will result in –100% return on investment.

Edge: This is the edge you have earned dependent on the cost and the revenue. If revenue is 0, the margin will be 0.00%.

6.16 Objective and Its Setting

Since we have a rough idea of how PPC functions, defining reasonable objectives to work viably with PPC and execution measurements is an unquestionable requirement [24]. This will empower us to distinguish "What we have to do" and "what is as of now accomplished" and if we see this is what marketing is all about! We need a measure to decide if we are getting our money's worth from the SEO way of promoting; this is the means by which objective and target setting encourages us. Prior to settling on them, we have to respond to a couple of targeted on, business-specific inquiries.

1) Do we need direct traffic, deals, or branding or a mix of all?
2) Do we have a lot of influencers with shared objectives?
3) Is there any negative effect that will be relieved?
4) Do we have the products and services built up on the web?
5) And likewise, we have to focus on below fundamental classifications:

1. **SEO for traffic:** This is to monetize the traffic without real exercises or monetary exchanges on the site. Focus on the specific keywords but only just to comprehend what is required; don't adjust the content totally to fit just the requests. Alter it, make it exceptional, and possibly focus on specific keywords when chipping away at titles of the content you have made. Additionally, focus on structure, in this manner making it simple to share and advanced for viral spreading.

2. **SEO for e-deals:** If you have a e-commerce website, this one ought to be in the objectives list and Basic guideline: The more explicit the inquiry, the more likely the guests are to make the buy.

3. **SEO for marking:** When we need an expansion of member participation, we should utilize this. Accomplishing a long sequence over your generally customary crude traffic is the objective here.

4. **SEO for reputation or goodwill management:** Now we know the effect of negative advertising, and that combined with the web is equal to a nightmare. Not at all like the different variations above, here the attention is more on utilizing social media profiles, influencers, official statements, and PR to adjust and push down current negative content.

6.17 Significant Metrics for Performance Measurements

Since we are finished with objective and target setting, we have to realize how we can measure our objectives and what we have accomplished to remain aware of the impacts or amount invested into SEO promoting. It is these measurements that help in basic leadership. They give us a chance to see a portion of the promotion estimation strategies with significant measurements:

1. **Distinguish Keyword "Exchange" Opportunities**

   Most web analytics tools dutifully report on search engine data, which search engine is sending how many visits and so forth. They additionally have a report for specific keywords. Also, unquestionably at this point you realize that you can bore down from the web index report to take a look at keywords for that web search tool. But in reality, the positioning of search engines by number of browsers to your webpage is very much settled. It has been for quite a while. It doesn't change month over month, or practically doesn't change. What exists in your information is option of how one keyword performs crosswise over web indexes! This is because of the distinction in calculations for both natural and paid pursuits in every motor. Arranging by clicks per specific keyword in another crawler, Yahoo! For this situation, we see that, among our top traffic terms, there are ones where Google is failing to meet expectations.

2. **Concentrate on "What's Changed"**

   While we fixate on our image terms and our best ten key expressions, actually the long assessment implies that our natural and search campaigns center around several thousand or a huge number of specific keywords. One viable procedure to manage this information issue is to concentrate on what's changed. No more information vomit. Simply take a gander at things that need consideration. The focal point of your What's Changed reports is to show ventures, advertisement forums, keywords that are producing more impressions, getting more clicks, and creating more income (or not!), contrasted with an earlier pertinent timespan. The most serious issue with paid pursuit, or web, investigation is that you don't have starting points. These reports, and measurements, give you that. You would then be able to go examine the hot leads. Maybe your rivals have gotten a move on. Perhaps your quality score has taken a jump. Who knows, the young men in the distribution center may have come up short on stock.

3. **Dissect Visual Impression Share, Compute Lost Revenue**

   It is extremely helpful in understanding what portion of shelf or stock rack you have. It is astounding that only the most no-nonsense PPC people appear to concentrate on this metric. You can also get the most click generating content share for your own specific keywords portfolio from the AdWords Campaign reports [25]. You need to know what number of individuals who are scanning for a keyword that you are offering on are not seeing your content advertisements. Winning those missing impressions would require either a spending increment, or, more probably, some huge improvement in offers or quality score – yet realizing that potential exists offers a clearer perspective on explicit development opportunity than paid browsers typically observe.

4. **Grasp the ROI Distribution Report (Identify: Lovers, Friends, Losers)**

Keep in mind the 80/20 guideline: 80% of ROI originates from 20% of your campaigns. Supplant ROI with income, benefit, your preferred measurement.

a. **Step 1**: Specify your ROI objective and your base satisfactory ROI.

b. **Step 2**: Run report – the report discloses to you how many of your campaigns, ad groups, and keywords fall into three execution groups: Great (surpasses desires), Good (meets desires), Poor (hurts!). Or on the other hand: Lovers, Friends, Losers. This kind of examination enables you to take a very close/basic look at precisely what number of your keywords are extremely beneficial, and which ones are forcing more expenditure. It should prompt a total reexamination of your keyword choices, coordinate sort settings, offer decisions, advertisement duplicate, and campaign association. This investigation is more accepted since you end up with an organized plan for the day of ventures or specific keywords that require your complete consideration.

5. **Focus in on the Actual User Search Query (and Match Type)**

While the whole business of paid search spins around keywords, frequently there isn't sufficient consideration being paid to the role that match types play in figuring out which search questions (the words the web crawler client really types) trigger your paid inquiry advertisement.

6. **This Investigation Will Instantly Spark Ideas For:**

a. Match type advancements (specific keywords and search inquiries that you are currently purchasing in phrase or broad match yet could and ought to be in exact match).

b. Search inquiries that are coordinating your broad match keywords yet should themselves be phrase match specific keywords.

c. A rundown of negative keywords that you should add to proficiently quit purchasing unfit traffic.

6.18 Detailing Account Structure

The accomplishment of web index advertising efforts rests, as it were, on their legitimate structure. An ill-advised SEM record structure undermines all other paid inquiry advertising endeavors, for example, improper budget management, troubles performing account enhancement, and less control of internet promoting endeavors. In this manner, it becomes critical to establish a strong framework from the very start and keep up it methodically as indicated by web best practices to maintain a strategic distance from inefficient spend and to drive deals or leads at a benefit. Sensible keyword gathering and record structure can enable you to accomplish higher click through rates, lower costs per click, and for the most part more grounded execution, and specific keyword research can enable you to consider how to best structure your record. AdWords and Bing Ads records ought to be organized in the following manner for ideal outcomes:

1. Ad campaigns
2. Ad gatherings or forums
3. Keywords
4. Ad content
5. Landing pages

Promotion ventures can, and ought to much of the time, center around comparative items or administrations.

6.19 Compelling Segmentation of Keywords and Usage of Multiple Match Types

To benefit as much as possible from SEM, "keywords explore" becomes helpful: It revolves around the idea of distinguishing the words identified with your item or administration that a client is probably going to utilize when paying special mind to your item/administration. Likewise, thinking about the negative keyword or unessential terms eliminate them from your content works to tailor the content well. In this way keyword choice incorporates match type choice and negative keywords. Wasteful aspects cause a distinction between the advertisement clicked and the capture term(s) entered. Once that user arrives on your site, however, and fails to discover what they are searching for because of the distinction, they will leave the site. *The most normal significant sorts of keyword match are* Wide, Modified Broad, Phrase, Exact, and Negative Match.

1. **AdWords Broad Match Type**: Wide match is the default match type and the one that contacts the most extensive group of spectators. When utilizing wide coordinates, your advertisement is qualified to show up at whatever point a client's pursuit inquiry incorporates any word in your key expression, in any order.

2. **AdWords Exact Match Type**: Precise match is the most explicit and prohibitive of the keyword match types. With this match type, visitors can possibly observe your advertisement when they type your accurate specific keyword express by itself. In terms of drawbacks, you will have less traffic because of your limitations, in light of the fact that these progressively explicit capture inquiries have lower search volume, and you won't get the same number of impressions. Match types can majorly affect your record's presentation: They're the control you use to decide precisely which search questions you're offering on. As you figure out which match types to use for every specific keyword, there are a couple of key segments to consider:

   a. **Performance to date**: How a specific keyword or comparable keywords have performed can give you bits of knowledge about which match type will give the best return on your venture.

   b. **Competitors**: How your rivals bid on specific terms and structure their own records, as well as how their records have performed truly, will all affect the traffic you see from certain match types.

   c. **Bids**: Cost per click and cost per conversation are intensely affected by offers – regularly publicists utilize different procedures for controlling offers and every now and again offer more or less aggressively dependent on the match type – this can unequivocally impact which match type is generally fitting.

   d. **Ad text and AdWords account structure**: Many sponsors will break out a "cash" keyword and run it on wide, expression, and careful match types – maybe even segmenting those keyword match types out and composing explicit promotions for each. The manner in which a publicist structures a record can likewise massively affect execution for various keyword match types.

6.20 Non-Overlapping Ad Groups

The ad group level holds the keywords and promotions that are destroyed to the web crawler results page when a web client enters a specific keyword or expression that is a focus of your web-based showcasing endeavors. At this point the account for PPC campaign must be clear. By making a record for our PPC campaign we continue ahead by making an ad venture, and this is trailed by ad bunches which resemble a holder lodging keyword, content advertisements, and presentation pages. In this manner, ad gatherings give a pecking order to everything.

6.20.1 For What Reason Do We Truly Require Ad Gatherings?

Ad gatherings or groups decide which specific keywords our advertisement will show up on, what our promotion states, and where it will direct the watcher to when clicked. In this way, basically it discloses to you whom to connect with your advertisement or rather gives you the correct audience and how might you stand out enough to be noticed.

Example: You wish to showcase your skincare items; you continue by making various campaigns for sorts of skin health management items (state summer healthy skin) and now you will concentrate on a single kind of item in an advertisement gathering (state sunscreen for Asian summers). In any case, don't make the mistake of thinking that you may have twofold the opportunity of progress by offering on a similar keyword more than once. Duplicate keywords in your PPC venture do not really help you over the long haul. AdWords will just indicate one case of a promotion for every specific keyword per publicist. So, in case you're focusing on the keyword expression "regular healthy skin", in more than one campaign, Google will pick the advertisement that is increasingly applicable or the better out of the two. Thus, if you are offering on a similar specific keyword as a similar promoter, at that point you're competing against yourself.

6.21 Characterize Performance Metrics

Since you are running an advertisement gathering, it is important to quantify it to capitalize on SEO or site design improvement ideas. Also, for this reason there are numerous measurements available. Consequently, an execution metric is "a quantifiable pointer used to survey how well your PPC venture is accomplishing its ideal targets".

Significant execution measurements are recorded underneath:

1. **Quality score**: Measure of the importance of your specific keywords.
2. **Click-through rate**: Low navigation rates are an indication that either your specific keywords or your ad creative need improvement.
3. **Conversion rate**: Discloses to you how many individuals who clicked your promotion proceeded to finish the ideal activity on your greeting page, obtaining, joining, or rounding out any structure.
4. **Cost per conversion.**
5. **Wasted spending cost**: Measures how much cash goes into paying for clicks that don't convert.

6.22 Screen PPC Activity with Google Analytics

Connecting Google Analytics to your AdWords record can enable you to break down client movement on your site after a promotion click or impression. This data can reveal insights into the amount of your site traffic that originates from AdWords, and help you improve the same [26]. Connecting records enables you to import Google Analytics objectives and exchanges, see Google Analytics information in your AdWords reports, and import analytics remarketing audiences. You'll additionally observe AdWords information in your analytics reports.

Google Analytics: A free device offering a vigorous usefulness with a great deal of information – the vast majority of which numerous entrepreneurs don't even use! You can chase around and discover your transformation data, which is tedious. Rather, simply set up your objectives and have Google order every piece of information for you. To screen the viability of your web-based advertising, focus on three noteworthy segments:

1. Create contact structures with affirmation pages.
2. Use an outsider call-following administration.
3. Set up objectives in the investigation stage.

6.23 Understanding Reports and Define the Plan of Action

We know how powerful web promotion is, and consequently it is vital that you comprehend traffic so that you spend your financial limit effectively. Google investigation reports can give you noteworthy data about what is and isn't improving your site and in your internet promoting endeavors [27]. Use Google Analytics or Yahoo Web Analytics to enable you to look at your SEO and PPC specific keywords. Focus on your ricochet rates since that is a significant marker in PPC. In the event that you are getting a high skip pace of over half, yet a high CTR rate (i.e. 3% or more), at that point you have to chip away at your greeting pages.

Inquiries to pose: Does your duplicate match the PPC promotion? Is your duplicate over the overlay? Is it clear what the client ought to do straightaway? If you are getting both a high skip rate and a low CTR rate – at that point you have to take a look at your PPC campaign. Take a gander at your ad message and ask yourself – does it coordinate the keyword query? You may require progressively negative specific keywords so you can get increasingly important traffic. A portion of the key parts of a Google investigation report are:

1. Page views: The occasions the page stacked.
2. Unique site visits: Separate visits to the site where that page is opened at least multiple times.
3. Visits: Are what might be compared to program sessions.
4. Bounce rate: Visits in which the individual left your site.
5. New sessions: First-time visits.
6. Goals: how frequently guest made a complete desired action.
7. Conversions: Number of times objectives were finished on your site.
8. Acquisition: How you secure visitors.
9. Behavior: Helps in improving content.

The different tabs help in modifying your report for the most basic areas. When you look at your Google investigation reports, you'll be keen on the traffic sources that send you the most guests. In any case, recollect that the volume of guests is just one of the measurements you ought to think about when assessing your traffic sources. The nature of that referred traffic matters as well.

6.24 Market Analysis and User Psychology

To help you attract, convince, and convert more people with your marketing, you should know the following user psychology [28].

1. **Priming**: Using subtle priming techniques, you could help your website visitors remember key information about your brand – and maybe even influence their buying behavior.
2. **Reciprocity**: If someone does something for you, you naturally will want to do something for them. In your marketing, there are a lot of ways to take advantage of reciprocity. You don't have to be rolling in dough to give something away; it can be anything from a branded sweatshirt, to an exclusive eBook, to a free desktop background, to your expertise on a difficult subject matter. Even something as simple as a hand-written note can go a long way in establishing reciprocity.
3. **Social proof**: One easy way to make the most of social proof is on your blog. If you're not already, use social sharing and follow buttons that display the number of followers your accounts have or the number of shares a piece of content has. If those numbers are front and center and you already have a few people sharing your

post, people who stumble on your post later will be much more likely to share.

4. **Decoy effect**: If you're looking to increase conversions on a landing page with two options, you might want to add a third. It could help increase the conversion rate of the option you'd ultimately want people to take.

5. **Scarcity**: The rarer the opportunity, content, or product is, the more valuable it is.

6. **Anchoring**: Anchoring is important to know – especially if you're ever running a sale. You'll want to clearly state the initial price of the product (this is "setting" the anchor), and then display the sale price right next to it. You might even explain how much of a percentage off your customers will receive with the sale.

7. **Clustering**: How can you design and lay out your content to increase memory retention? One way to do it is by grouping similar topics together – either under numbered bullet points or with different header sizes. Besides being much easier to scan, your writing will be much easier to remember and recall down the road – especially if you're creating an interesting advert.

6.25 Understanding Industry Key Drivers

There is an assortment of inbound promoting systems, yet we should begin with the five key drivers of site traffic. Consider them the fundamental elements for your showcasing formula.

1. **Site Design Improvement by SEO**

   Having an excellent site and incredible showcasing content has constrained worth unless your site can be found and the content can be associated with ROI [29]. It's basic to have a content management system where you or your organization can consistently screen what's going on. Old search engine optimization strategies like stuffing keywords into meta information and building huge volumes of inbound connects to increase what was once observed as "authority" – no longer help with organic search results.

2. **Web-Based Social Networking**

   Web-based social networking is currently considered the number one driver of all site referral traffic [30]. Adopting a functioning strategy to advancing your site content by means of social promoting is basic. Facebook, Twitter, and LinkedIn now give numerous alternatives to focus on and advance publicizing. Moreover, Facebook and

Twitter offer remarketing, which is extraordinary for B2B and B2C advertising. By essentially introducing a following pixel on your site, you can keep on promoting to visitors after they've visited your site, which is an incredible method to encourage return traffic to your bank or credit association.

3. **E-Mail Marketing**

E-mail advertising is a ground-breaking strategy for arriving at your existing client base. E-mail segmentation is incredible for conveying custom messages that are important to your group of spectators [31]. Consistency is key with e-mail advertising. That is the reason monthly pamphlets and customer releases are so effective. Endorsers start anticipating new content all the time.

4. **Search Engine Marketing (SEM)**

SEM is paid promotion inside the Google and Bing web crawlers. Both give you immediate site traffic. Google AdWords is the most broadly utilized compensation-per-click model, where visitors can bid on keywords and pay for each click on their notices. You can target explicit geographic regions and areas and select keywords explicit to your contributions and administrations to direct people to your site [32].

5. **Blogging**

Blogging is the foundation of effective traffic building. Creating and sharing important, rich content can help with your web crawler results. Constant routine with regards to these systems will help drive more traffic, yet that is just a piece of the story. To effectively convey a "high volume of qualified guests", the content on your site should seriously connect with your crowd. This will guarantee they are the correct kind of guests and that they return [14].

6. **Focused Competitive Analysis**

SEM benefits organizations in increasing presence over major search engines and provides another channel to promote the brand and company. This additionally enables an association to talk legitimately, straightforwardly, and sincerely to its visitors. SEM helps in keeping up aggressiveness in the market, empowering an association to remain on top [16].

Steps for fruitful comprehension and analysis some opportunities available Customer satisfaction is paramount.

1. Product and administration inclinations and purchasing conduct investigations.
2. Exploration of social and attitudinal inspirations.
3. Lifestyle investigations.

Furthermore, we as a whole are in business so competitor investigation is an unquestionable requirement: Read ahead to know the primary Ws when examining a rival.

1. Who are your rivals?
2. What sort of contenders would they say they are?
3. Where would they say they are rivaling you?
4. Why would they say they are competing?
5. When would it be advisable for you to make a move?

With below flow of analysis, you will see how to find a response to the above questions.

Step #1: Pull the information!

Get acquainted with AdWords and Bing Ads' Auction Insights, which can be gotten to through the "Subtleties" drop down in both stages. The measurements you'll need to take a gander at are:

1. **Impression share**: How frequently another sponsor got an impression, as a proportion of the sales for which you were additionally contending [18]. Note that your impression offer is an estimation of the quantity of impressions you got divided by the evaluated number of impressions you were qualified to get.
2. **Cover rate**: How frequently another publicist's promotion got an impression when your advertisement likewise got an impression.
3. **Position above rate**: How frequently the other publicist's promotion appeared in a higher situation than yours, when both of your advertisements appeared simultaneously.
4. **Top of page rate**: How frequently your or your rival's advertisement appeared at the highest point of the page, over the natural list items.
5. **Outranking share**: How frequently your promotion positioned higher in the sale than another promoter's advertisement, or your advertisement appeared when theirs didn't.

Step #2: Take the information and make it significant

By pulling segmented campaign and Auction Insights information and filling it into a Google Sheets doc, you can automatically produce graphs that represent focused competitive metrics alongside your trending cost per click (CPC).

Step # 3: Are you going up against:

1. **Affiliates**: Marketing accomplices that utilize an outsider connection coordinating to your brand site to gain commissions on deals.
2. **Comparison shopping engine (CSE)**: An engine that lists products that send to outbound links for your retailer site.
3. **Online travel office (OTA)**: A webpage where searchers can book travel-related services.
4. **Partner**: Marketing accomplice who isn't a subsidiary and advances your brand's item or administrations.
5. **Search arbitrager**: Site that attempts to pull in traffic to a page that demonstrates extra advertisements.
6. **Reseller**: Retailer that sells your brand's items or administrations.
7. **No trademark**: Advertiser who doesn't have authorization to utilize your trademark, however is doing as such.

Watch out for which specific keywords your SEM subsidiaries, accomplices, or affirmed filiates are offering on. Have discussions on specific keyword arrangement early to guarantee you're not pointlessly blowing up CPCs. Track impression share week by week to recognize any real vacillations.

Step #4: Determine whether you're getting the full story

PPC sell-offs aren't in every case obvious. In case your accomplices or associates are pointing to a similar domain as you, at that point Google and Bing won't have the option to separate you both in Auction Insights and will lump you both under a similar area (recorded as "You"). Overseeing affiliates versus non-endorsed trademark visitors can likewise be troublesome (particularly if you have a large number of affiliates). Keep in mind that it's additionally challenging for the approved resellers on the other side of the equation.

Step #5: Set up alarms, reports, and robotized offering rules

Set up robotized Search Impression Share reports for top performing efforts, advertisement gatherings, or keywords to send to your e-mail. Use Google and Bing's robotized guidelines or Google's adaptable offering procedures like Target Search Page Location to shield keywords from slipping from places that result in the best return. A significant number of your rivals will attempt to demonstrate advertisements on your specific keywords for similar reasons as you, and you won't have the option to beat them with simply offering alone. Set aside the effort to audit your rival's system on the open SERP and build up an activity plan around your learnings.

Step #6: Review your rivals' messaging regularly

We all know how to search for keywords legitimately on a given web index. Be that as it may, this technique prompts expanding impacts on your advertisements, messing with CTR. To get a feeling for what the SERP looks like continuously, you can utilize the Ad Preview and Diagnosis Tool as a beginning stage. However, in case you're searching for more detailed information on your rivals' promotion duplicate, there's a significant number of apparatuses to help you from SpyFu, iSpionage, and SEMRush.

Step #7: Test your advertisement duplicate methodology

You don't get the opportunity to control precisely what your rival places in their promotions; however you can test your own messaging. What are your rivals advertising? Is it true that you are utilizing clear suggestions to take action? In case you're in a jam-packed vertical, how might you stand out from different advertisements? How regularly would you say you are thinking about your rival's procedure when composing your duplicate for advertisement testing? Try not to spend each waking minute pondering your paid search competitors. You first ought to drive results for your record. Yet, contenders can shake your pontoon, and realizing how to control the ship through that is first step in the right direction.

6.26 Authoritative Positioning and Targeting

SEM gives a preferred position to contact previously motivated visitors. The segment(s) or group(s) of individuals and associations you choose to offer to is known as an objective market. *Directed showcasing, or separated advertising,* implies that you may separate some part of showcasing (offering, advancement, cost) for various gatherings of visitors. It is a moderately new marvel. Mass promoting, or *undifferentiated advertising,* started things out. It advanced alongside large-scale manufacturing and involves offering a similar item to everyone. You can consider mass showcasing as a shotgun approach: You can make however many advertising messages as would be prudent on each medium accessible as frequently as you can bear. To capitalize on SEM, targeted publicizing is the correct approach. While adopting this strategy you can likewise profit in the following ways during the division and focusing on procedure:

1. Avoid head-on rivalry with different firms attempting to catch similar visitors.
2. Develop new contributions and expand profitable brands and product lines.

3. Remarket more established, less beneficial items and brands.
4. Identify early adopters.
5. Redistribute cash and deals endeavors to concentrate on your most gainful visitors.
6. Retain "in danger" visitors in risk of deserting to your rivals.

Making sense of "who's who" in terms of your customers involves some analyst work, however – regularly statistical surveying. An assortment of instruments and research procedures can be utilized to section markets. Innovation is likewise making it simpler for even little organizations and business visionaries to assemble data about potential visitors. With the expanded utilization of web-based social networking, organizations can get data on customers' search behavior. Loyalty cards that shoppers scan at numerous basic food item and medication stores give an unbelievable measure of data on purchasers' purchasing conduct. Organizations are presently utilizing the internet to track individuals' web-perusing examples and section them into objective gatherings. Indeed, even independent companies can do this cost effectively on the grounds that they needn't bother with their own product and projects. They can essentially sign up online for products like Google's AdSense and AdWords programs. You can find potential visitors by looking at web journal sites and discussions on the web [15].

Finding new visitors, becoming more acquainted with them, and making sense of what they truly need is additionally a troublesome procedure, one that is full of experimentation. That is the reason it's so critical to become acquainted with, structure cozy connections, and focus selling efforts on current customers. In addition to studying their buying patterns, firms additionally attempt to improve comprehension of their visitors by surveying or contracting advertising examination firms to do as such or by using steadfastness programs. Notwithstanding how well organizations know their visitors, recollect that a few visitors are profoundly gainful, others aren't, and others wind up costing your firm cash to serve. Therefore, you will need to collaborate with certain visitors more than others. In all honesty, a few firms purposely "un target" unbeneficial visitors [16].

Steps engaged with target showcasing:

1. **Establish transient measures to assess your endeavors**. Decide how you will gauge your exertion. Will you utilize higher consumer loyalty appraisals, expanded incomes earned per client, number of items offered to visitors, exchange costs, or another measure?
2. **Identify your visitors**. Accumulate all the data you can about your present visitors, including their purchasing behaviors, likes, and abhorrence. When conducting business with them, incorporate "opt in" question that enables you to legitimately assemble and utilize

their telephone numbers and e-mail addresses so you can stay in contact with them.

3. **Differentiate among your visitors.** Figure out who your best visitors are as far as what they spend and will spend later on (their client lifetime worth), and how simple or troublesome they are to serve. Distinguish and target visitors that go through just limited quantities with you yet huge sums with your rivals.

4. **Interact with your visitors, focusing on your best ones.** Discover ways and media in which to converse with visitors about themes they're keen on and appreciate. Spend the main part of your assets collaborating with your best (high-esteem) visitors. Limit the time and cash you spend on low-esteem visitors with low development potential.

5. **Customize your items and promoting messages to address their issues.** Attempt to redo your showcasing messages and items to give your visitors precisely what they need – regardless of whether it's simply the item, its bundling, conveyance, or the administrations related with it.

6.27 Comprehend and Connect with the User

Customer identification is subsequently the core of everything in SEM. A few means have been recorded below:

1. **Open the alternatives.** Any individual who ventures online realizes protection issues exist, yet despite everything they go on the web and approve of it, realizing organizations are utilizing it to convey a profoundly customized way to deal with every client. Offering visitors a decision is the initial phase in building trust and certainty. Utilizing centered information and data to exhibit those choices to the client better positions an organization to keep up or add to its client base [17].

2. **Use information for knowledge and maintenance.** Advertisers will in general center fundamentally around the procurement, however they ought to likewise be centered around development and maintenance.

3. **Concentrate on continuous prospects, not future prospects.** Keep your mind in the present with an eye toward what's to come. Gather promptly accessible business records and other information in order to size up potential customers and crafting their approach.

4. **Contract sales reps to fabricate exceptionally targeted relationships.** Make your sales substantially more much more effective at their jobs. Truly, they burn through so much time doing research behind a desk and attempting to teach themselves about organizations and customer interests before connecting.

5. **Benefit from search behavior of forthcoming client.** Give us initially a chance to talk about how to inquire about the planned client. Use Google Alerts, Mention, or Talkwalker Alerts. Whichever alarm service(s) you're utilizing, look at the results to be certain your pursuit parameters are set accurately. If your prospect is an organization traded on an open market or an independent company dynamic in their nearby network, you will rapidly find out about their new activities, premiums, and exercises. If you are attempting to screen the conduct of tech purchasers, for instance, a Google Alert will tell you at whatever point new research is distributed.

6. **Meeting current visitors.** This may appear glaringly evident, however when was the last time you conversed with your visitors at any length? They are an ideal asset since they've acquired your item or administration and are moderately open to you. Meeting visitors won't just give you knowledge into their basic leadership process, it will likewise be an incredible chance to assemble content for a contextual analysis. Offering to get ready and advance a joint contextual analysis can be a win for both you and your client. Notwithstanding conversing with visitors one-on-one, you can likewise consider reviewing prospects or directing focus groups. (The subject of a future blog entry without a doubt.) You will likely recognize common interests, data sources, and difficulties.

7. **Concentrate your web investigation.** There's a huge amount of information available to you through your web examination, yet would you say you are utilizing it to become familiar with your purchasers and others with shared qualities and interests? Ask yourself inquiries, for example,

   a. What are the examples of guest conduct?

   b. Where do they originate from?

   c. What keywords did they use to discover you?

   d. Where do they go while on the site?

   e. How long do they remain?

   f. What content organizations are generally well known?

Do these examples reveal to you anything about where your visitors are in their purchasing procedure, or what content is best at the various phases of their purchasing procedure?

1. **Utilize your rivals.** Not exclusively do your visitors approach more data than at any time in recent memory, but you do as well! One way to get great insights into your buyers is to think about the examination or contextual investigations that your rivals have distributed. Looking into their contextual analyses may enable you to better comprehend your prospect, for example, why they may have picked your rival over you previously. In addition to following the competition, pursue industry expert web journals and reports.

2. **Influence proficient informal organizations.** Utilize the significant expert systems (like LinkedIn and Quora), and attempt to discover other, industry-explicit systems where your prospects may be. These systems are an extraordinary opportunity for listening and engaging. They will enable you to all the more likely comprehend the day by day difficulties or victories your prospects have, and much of the time, offer you the chance to pose inquiries of that network and get genuine, keen reactions. These systems are additionally an extraordinary method to upgrade the information you as of now have about individuals, and perceive how they are associated with other individuals inside their own associations or whom you know.

3. **This exploration helps in affecting the client in the following ways.** Apart from the content of your site (which buyers likely acknowledge is one-sided toward whatever item or administration you're offering), there are different things that will probably come up in indexed lists that affect whether they work with you. Some of them are as follows:

   a. **Online reviews**: Search any noticeable business on the web and you will undoubtedly observe Google Reviews as one of the top outcomes. Apart from Google, other survey sites like Angie's List and Yelp can either urge your potential client to buy from you or deter him altogether. By checking your organization's online surveys and reacting to them in an auspicious way, you can prevent the loss of potential new visitors [27].

   b. **Competitor websites**: If the site of an organization like yours reliably positions higher in query items, there's a decent chance you're losing numerous imminent visitors to the challenge. The most ideal approach to outrank your opposition is to contemplate what they're progressing nicely, and plan your site content to be far better [26].

   c. **False information**: Regardless of whether false data about you shows up on trick sites, websites, or different malignant sites, these can contrarily affect your business's notoriety and lead a client to choose one of the numerous different alternatives accessible to them on the web [28]. For a circumstance like this, you can hire a PR company to help, or you can take matters into your own hands. At times, you can have false data expelled from sites if it abuses that site's terms of administration.

6.28 Promotion Writing Techniques for Advertisement

Extraordinary promotion features resemble the conspicuous showcases you find in store windows consistently. They're there to leave you speechless, make you imagine yourself owning whatever it is they're selling, and empower you to cross the mental limit and stroll into the store to get it [22].

1. **Incorporate keywords**: Try not to make your prospects confused about what you sell. Make the association between their pursuit inquiry and your advertisement perfectly clear by incorporating the specific keyword in the feature.

2. **Pose inquiries**: Leveraging user intent is crucial to increasing transformation rates. One approach to do this is by asking the searcher an inquiry with your feature.

3. **Tackle prospects' problems**: This system is similarly as vital with regards to your promotion features for what it's worth in the body duplicate of your advertisement. Individuals would prefer not to purchase "things" – they need to tackle their issues.

4. **Include a little humor**: Sponsors frequently give close consideration to the kinds of advertisements their rivals are running. Shockingly, prospects frequently don't – all they see are many promotions that all appear to be identical. Separating your promotions from your rivals' is essential, and diversion can be a magnificent method to achieve this.

5. **Incorporate numbers or statistics**: Numerous visitors react well to exact proof. Hard information can be a sign of trust, and it can plant the seed of reliability in the psyche of your prospect.

6. **Contemplate user intent**: Keep in mind, no one thinks about you or your item – just how it can take care of their issues. Obviously, figuring out what those issues are means considering thinking about the search that showed your promotion to the prospect in any case.

7. **Use empathy**: Just as they need to take care of their issues, individuals need to realize that another person comprehends what they're experiencing. This is the thing that makes sympathy such an amazing system in the best promotion features. By identifying with your

client's concern, you're making a bond between you, which can build trust – or, in any event, grab their attention.

8. **Utilize simple language**: The digitalized advertising industry has a greater number of popular expressions and language than a significant number of us care for, and I'm certain you most likely utilize some particular terms in your industry, as well. In any case, that doesn't mean you ought to pack your PPC promotion features with enough language to befuddle or discourage your prospects.

9. **Utilize social trends as inspiration**: See what individuals are discussing in your industry through internet-based life and utilize that to inspire your headlines. Twitter offers a supportive "What's trending" area to help you rapidly recognize trending themes dependent on your inclinations, and jumping on an impermanent temporary fad could enable you to gain by social patterns to catch extra clicks.

10. **Utilize the character limit**: Try not to utilize superfluous words or characters for their own sake, yet ensure you exploit the 25-character limit in your PPC advertisement features. Be as distinct as possible, and ensure that your promotions plot precisely what visitors can expect when they click on them.

11. **Try not to make false promises or bogus claims**: Including explicitly false data is a quick way to have your promotion opposed by Google and Bing. If you make a case in the feature of your promotion, ensure either the advertisement duplicate or the greeting page (ideally both) back it up with genuine information, client tributes, or some other certain proof.

12. **Be obsessive about punctuation, spelling, and grammar**: Just because you just have 25 characters to work with in your PPC advertisement features doesn't mean you shouldn't invest energy ensuring your promotion is syntactically right. Incorrect spellings will make you and your business look absurd, so run everything through a spellchecker before presenting your advertisement for endorsement.

13. **Concentrate on the benefits**: Keep in mind that individuals need to realize how picking your item or administration will profit them. Deals experts regularly remember a mantra – "How might this benefit me?"

14. **Keep an eye on the competition and steal their ideas**: Corporate surveillance is extraordinary for business! Luckily, you don't have to break into the corporate base camp of your greatest opponent or hack into their PC frameworks to perceive what sorts of promotions are working for them.

15. **Compelling ads that increase click through rate (CTR) lower costs**: CTR is a key estimation that Google uses to assess the congruity among keywords and advertisements [21]. Essentially, it addresses what number of individuals are tuning into your offer versus how

many are thoroughly disregarding you. Brilliant guidelines to improve it:

a. Write better promotions: Aim to give a response to the searcher's inquiry.

b. Use Sitelink Ad augmentations: Sitelinks give an approach to advance related administrations and assets that supplement your core service offering.

c. Extend headline with your first description: For example: Natural healthy skin – 30% off on summer special skin care products. Here the significant point is how your augmentation of a heading works in attracting traffic.

d. Use prohibitive match type.

e. Manage geography: Down bid or reject districts with low CTR – investigate how your administration performs in various sites of the nation and settle on this choice.

6.29 Understanding, Analyzing, and Improving: Relevance and Quality Score

Quality score is a number dictated by Google that rates the general quality and significance of your AdWords advertisements and keywords. The situation of your advertisement on the list items page, and the measure of your expense per click depend on this number. Subsequently, the Quality Score is influenced by expected CTR, advertisement importance, and point of arrival experience. What's more, in the event that we talk business: The better your promotions address the issues of a searcher's solicitation, the higher your quality score will be and the less you'll need to pay for each click.

6.29.1 Enhancing Quality Score

Here's a great list of best practices for accelerating your website. The quality of the website can be enhanced by following the seven practices listed here.

1. **Increase your CTR**: A high CTR (active clicking factor) implies that visitors are tapping on your advertisement when their pursuit inquiry matches with one of your specific keywords. Expanding your CTR will prompt a lower cost per click (CPC). In the event that you have a low CTR, Google will give you a low Quality Score, which prompts expanded expenses over the long haul.

2. **Analyze your pursuit impressions share information**: It's critical to break down pursuit impression share information. Impression share (IS) is the quantity of impressions you've received, divided

by identification number of impressions you were qualified to get. Qualification depends on your present advertisements' focusing on settings, endorsement statuses, offers, and quality. If your inquiry impression offer is low, you can build it in two different ways: (a) By expanding your everyday spending plans or boosting offers to rank in higher positions, and (b) by utilizing wide match changed keywords (utilizing specific keywords in accurate and expression match will bring about moderate impression development).

3. **Group specific keywords in advertisement gatherings to improve quality score**: Organizing your campaigns into smaller yet targeted ad groups is pivotal for a decent Quality Score. By following this training, you're expanding relevancy between the inquiry question and the promotion. Try not to have only a couple of promotion bunches in your campaigns. After beginning the investigation of specific keywords, take a gander at your keywords rundown and gather them into sensible gatherings or topics. This will assist you with making your advertisement gatherings engaged and pertinent. Numerous B2B organizations neglect to pursue this training and wind up paying more than the individuals who do see how Quality Score truly functions.

4. **Write top-notch content promotions**: Composing top-notch content advertisements can be challenging because of all the information you should fit into a small amount of space. Ensure you pick the advertisement duplicate that is firmly identified with your keywords to expand the pertinence of your specific keywords to every promotion. Expanding the pertinence will build your Quality Score. You additionally need to add your keywords to the advertisement duplicate. There are two different ways to do that: By physically adding keywords to your advertisements or by utilizing specific dynamic keywords inclusion (DKI). Dynamic keywords inclusion choice will embed specific keywords naturally in your advertisement duplicate.

5. **Check your greeting page burden speed**: Presentation page burden speed has turned into a significant factor in ascertaining Quality Score. Invest some energy dissecting your page burden speed. Web visitors have progressively limited capacity to focus, and increasingly more frequently, we see visitors tapping on advertisements and skipping even before the page has stacked. To keep that from occurring, improve your points of arrival to diminish their heap speed however much as could be expected. Perhaps the most straightforward approaches to improving your heap speed is to have your site in the area you serve your visitors. As it were, if your visitors are US- based, don't have your site in the UK. Other than that, there are

numerous ways you can upgrade your point of arrival. Besides that, there are many ways you can optimize your landing page.

6. **Keep an eye on your skip rates:** There is no documentation from Google that says skip rate can diminish your Quality Score, albeit some have estimated this might be valid [20–23]. While bob rates don't legitimately influence your QS, high skip rates show there is a major issue with your keywords, advertisements, or presentation page. At the point when that occurs, ensure you're passing on the correct message on your greeting page and your advertisements. This could mean changing one of them or both.

7. **Consider a Quality Score tracker script:** Before proceeding to next part, we'd like to impart one additional tip to you. To follow quality score changes after some time and have them recorded, we propose you utilize a Quality Score tracker script [21–23]. It's anything but difficult to actualize and fantastically important. In the next part, you can discover more data on the best way to set it up. Google doesn't store recorded Quality Score information in your records, so this content will assist you with figuring out how your Quality score changes over time.

6.30 Important Landing Pages

A point of arrival is made for a solitary change objective, for example, a whitepaper download or e-mail bulletin link, or to call and make an arrangement. Each Google promotion you make should prompt a particular point of arrival that was structured and written to complete the conversion after somebody taps on the advertisement. Here are ten hints for making compelling points of arrival for your next Google AdWords campaign:

1. **Coordinate the keywords, ad, and landing page:** To see the best outcomes from your computerized promoting effort, you need a system that keeps all parts of it joined together. Another point of arrival must be explicitly created for every advertisement, and it should utilize similar keywords and related content to that used in the promotion itself.

2. **Use text to win over site visitors:** Improving your active visitor clicking percentage (when a browser clicks on the advertisement to go through to your landing page, otherwise known as CTR) doesn't ensure an uptick in transformations. The content on your advertisement's goal page should outline your special selling focuses, urging

customers to pick you over a contender. Otherwise, they may basically hit "back" and visit a contender's site.

3. **Make the site experience trustworthy**: The content on a presentation page isn't the main thing that can work for you, in any case. A type of visual content can help give your page more body, making it look increasingly like a genuine page and less like a quick sales ploy. Consider including pictures of your business or group, installing a short video, or including a diagram or infographic that rapidly outlines data.

4. **Make each landing page different**: This tip is especially significant for local business organizations, for example, small health care practices that take into account nearby customers. You can run a promotion campaign that targets shoppers in various states, urban areas, or rural areas. Every advertisement, since it will utilize an alternate geo-tag in its keyword expression, should prompt an extraordinary presentation page.

5. **Avoid the temptation to cut and paste the same content from page to page as Google punishes you for this**: Besides, you ought to incorporate a couple of expressions and focuses that let purchasers realize you are talking legitimately to individuals like them, in their neighborhood, and not simply to a mass-advertised group of spectators.

6. **Test ad groups**: No matter what number of articles about computerized publicizing you read, it is difficult to tell precisely how well Google AdWords will function for you without testing it. You can run variations of a similar promotion simultaneously to see which one gets more clicks. You can likewise run a similar promotion, however have it lead to various points of arrival to see which goal page performs better with lead transformation.

7. **Include contact information**: You can track click-throughs and lead transformations through Google AdWords; however remember that numerous purchasers may change over in a totally unique manner. They may round out a short online structure to pose an inquiry or solicit data, or they may just pick up the telephone and call your office. Including your telephone number is a top need, however other ideas include the physical location, e-mail address, and direct links to online networking profiles. Medicinal services practices may likewise need to incorporate a little guide or Google Maps module that rapidly shows potential customers your area.

8. **Keep the call to action near the top**: If your promotion will probably build e-mail pamphlet information exchanges, at that point don't make customers look down to discover the information structure

[18]. Keep your CTA "over the overlay", in a manner of speaking, and make it simple to see so you don't lose potential changes. To be completely forthright, this applies to your landing page also.

9. **Appeal to the "skimmers" among us**: Once more, a point of arrival is altogether different from a blog entry or on-location article. Customers are not there to find out about another item or indicative advancements; they are there in light of the fact that they navigated from a promotion. Continue posting extraordinary content all through your site, yet keep the greeting page itself basic and clean. Appeal to busy individuals who are increasingly well-suited to skim and sweep content for the data they need at the time by utilizing visual cues or numbering frameworks.

10. **Use meta tags and title tags**: Meta and title labels are the foundations of SEO, which means they are likewise fundamental for any Google AdWords campaign. The web index creeps and records pages, recovering the most fundamental data to shape a synopsis of each page [4]. The title label alludes to the page's title, which ought to be brief and incorporate keywords. The meta labels allude to the chosen set of keywords, headers, site portrayals, and alt labels that contain more data about the page.

11. **Pay attention to the landing page URL**: Each page on your site has an unmistakable URL, which is the thing that shows up in the location bar of the program. Most buyers don't peruse them; however, Google does. Indeed, Google AdWords will utilize the URL to recommend which keywords could be utilized in the advertisement. Ensure the words you pick are consistent along with your keywords and transformation.

6.31 Significance of UI/UX Design

UI represents user interface, which incorporates the screens, its pages, buttons, and all other visual components that are utilized to enable a client to communicate with a device. UX alludes to refers to the user experience design. It incorporates the general experience that a client has as they collaborate with a site. This can incorporate the UI; however it can likewise consolidate significantly more. It is tied in with getting somebody from point A to point B in an apparently easy process on the client's side. In this manner we see UI/UX planners explore them through your webpage in the most effective manner conceivable so they can get the item or administration they need, and that implies the best sites/applications are those that react rapidly and proficiently.

6.31.1 UI/UX Configuration Tips

1. **Know Your Visitors**

   That implies knowing all the statistic information your examination app(s) can pull, yes. In any case, more significantly, it means recognizing what visitors need, and what disrupts the general flow of them accomplishing their goals. Getting to that degree of sympathy requires more than cautious examination of details. It requires becoming more acquainted with the individuals who utilize your site. It means talking with them up close and personal, watching them utilize your item (and possibly others), and asking them inquiries that go further than, "What's your opinion of this design?" Don't stop at knowing what your users want. Dig deeper and find out what they need. All things considered, wants are only outgrowths of necessities. If you can address a user's deep-seated need, you'll address their wants while also fulfilling more fundamental requirements[19].

2. **Characterize How Individuals Utilize Your Interface**

   Before you plan your interface, you have to characterize how individuals will utilize it. With the expanding pervasiveness of touch-based devices, it's a more urgent worry than you may think. People use sites and applications in two different ways: Straightforwardly (by associating with a component of the item) and by implication (by connecting with a component external to the product). Who your visitors are and what devices they use ought to profoundly illuminate your choices here. In case you're structuring for seniors or others with constrained manual skill, you wouldn't have any desire to incline toward swiping. In case you're planning for essayists or coders, who principally associate with applications through the console, you'll need to support all the common keyboard shortcuts limit time working with the mouse.

3. **Ponder Component Arrangement and Size**

   The closer and additionally greater something is, the quicker you can put your cursor (or finger) on it. This clearly has a wide range of suggestions for communication and UI structure, yet three of the most significant are: Make catches and other "click targets" (like symbols and content connections) big enough to effortlessly observe and click. This is particularly significant with menus and other connection records, as deficient space will leave individuals tapping an inappropriate connection over and over. Make the buttons for the most widely recognized activities bigger and increasingly conspicuous. While you're pondering component setting and size, consistently remember your collaboration model. If your site requires horizontal scrolling rather than vertical scrolling, you'll need to consider where and how to cue users to this unusual interaction type.

4. **Make Your Interfaces Simple to Learn**

   Limit the number of things an individual needs to make sure to utilize your interface productively and adequately. You can encourage this by lumping data together, i.e., breaking it into little, absorbable pieces. This thought dovetails with Tesler's Law of Conservation of Complexity, which expresses that UI architects should make their interfaces as straightforward as could be expected under the circumstances. That can mean concealing the unpredictability of an application behind an improved interface at whatever point conceivable. A famous case of an item neglecting to adhere to this law is Microsoft Word.

5. **Settle on Basic Leadership Straightforward**

   Such a large amount of the web shouts at us: "Banners" all of a sudden grow to turn out to be full-screen promotions. Modal boxes spring up, entreating us to buy in to online journals we haven't got an opportunity to peruse yet.

6.32 Source of Inspiration

Definition: "a bit of content proposed to instigate a watcher, visitor, or audience to perform a particular demonstration, ordinarily appearing as a guidance or order". Along these lines, it alludes to any device intended to incite a quick reaction or energize a prompt deal. A call to action (CTA) regularly alludes to the utilization of words or expressions that can be consolidated into sales scripts, publicizing messages, or site pages that urge purchasers to make a quick move. Consequently while taking a shot at CTA you ought to think about offering something of significant worth like a discount, animate interest, utilize social confirmation by expressing the individuals who have just tapped on CTA model "See why eight out of ten pick us over our rivals", include courses of events "substantially valid for seven days" and so on.
CTAs ought to be:

1. Visually impactful with copy that compels you to click the offer.
2. Brief: A few words is ideal, close to five is perfect.
3. Action-situated: Begin with an action word like "Download" or "Register".
4. Located in a simple-to-discover spot that follows naturally from the progression of the site page.

5. In a differentiating shade from the shading plan of the website page, while still fitting in with the general plan.

6. Large enough to see from a distance, however not all that enormous as to diminish consideration of the primary content on the page.

7. Easy to comprehend and clear: Be certain to state precisely what the guest will get in the event that they click on the CTA and go to the presentation page.

Presently let us see how CTA advantages us:

1. **CTAs to produce leads**: The motivation behind this CTA is to pull in guests and transform them into leads; in this manner, you'll need to put a CTA in an area on your site that sees countless new guests, as on your blog. The best places you could put a CTA are toward the end of a post, in the sidebar, and as a drifting standard in the corner [20].

2. **Social sharing CTA**: It might appear glaringly evident, however social sharing catches, as for Twitter and Facebook, are a successful, simple path for guests to wind up associated with your organization. Spot these CTAs on blog entries and points of arrival, yet be mindful so as to exclude them in spots where individuals are presenting their own data.

3. **Lead nurturing CTA**: You need to further urge visitors to purchase your item, so you should initially entice them with another offer that relates legitimately to your item, for example, an item demo or free preliminary. Spot these CTAs in key spots where leads will in general reliably visit. You could even utilize a shrewd CTA in a blog entry or on another offer's thank you page.

4. **Close the sale**: Utilizing a CTA is a decent strategy to transform leads into visitors. The motivation behind this CTA is to urge a potential client to purchase your item or administration directly right then and there when they are discussing whether to tap on the CTA. Make potential visitors feel good while engaging with your organization through making a well-disposed and clear CTA. You can put this sort of CTA on an item page and you can likewise utilize shrewd CTAs toward the end of blog entries.

5. **Promote an advantageous occasion**: Utilize an event which is advantageous for promotion CTA to attract the attention of and pull in guests to your occasion. How much simpler would it be to get the opportunity to find out about and register for an occasion? The area alternatives on your site are vast for this CTA since the message can identify with all extraordinary spectator types.

6.33 Advertisement Preview and Other Advanced Tools

An advertisement preview device shows you a review of a Google list items page (where you can decide whether your promotion is appearing or not in the wake of entering an inquiry term), yet in addition you can perceive how your advertisement really shows up in various devices and land areas without producing a solitary impression. Google these days is giving us individual outcomes, so if you have had some particular enthusiasm for your rival's area in the most recent few months, at that point his space may indicate higher when you search, however not for customary searchers. To keep away from perplexity and to have the option to see the precise advertisement position and condition of your promotions, it's best prescribed to utilize the Google AdWords Ad Preview Tool [21]. With this determination instrument, you can find:

1. If your promotion is being shown or not.
2. The name of the promotion gathering and venture that contained the activated keyword.
3. The keyword that sets off your promotion and its match type.
4. Reason why a specific keyword isn't being activated.
5. If you click where it says "See 2 additional keywords", you will see a similar specific keyword with various match types and the reason why the promotion isn't appearing with that particular match type.

The Ad Promotion Preview and Diagnosis Tool have a few helpful highlights that you should exploit. The Diagnosis Tool only works if you are logged into your AdWords account. This is on the grounds that it takes the data from the AdWords record to restore the name of the campaign and the name of the promotion bunch that contains the keyword that set off your advertisement.

6.33.1 Benefits of Utilizing This Device

1. You can check if your advertisement is appearing and become familiar with its situation without influencing the promotion's presentation.
2. You can perceive how your promotion appears on different changed devices.
3. It enables you to choose various areas, dialects, devices, and Google spaces, to coordinate your focused-on crowd. This is particularly helpful in case you're focusing on other geographical zones than where you dwell.

4. With the Diagnosis Tool you can see the specific keyword that triggers your promotion, the name of the campaign, and the advertisement bunch that contains such keywords.

5. The Diagnosis Tool can enable you to figure out why your advertisement isn't appearing (e.g. low promotion rank, low offers, low Quality Score, copy specific keywords, negative keywords).

6.33.2 Imagine a Scenario in Which Your Promotion Isn't Appearing

In the event that your promotion isn't appearing in the Ad Preview Tool, at that point I would propose that you look all the more carefully at the accompanying potential components:

1. Is the everyday spending plan depleted for the afternoon?

2. Check your advertisement's status and ensure it's empowered.

3. For Sitelinks that are not appearing, recall that your promotion should be in one of the top places of Google's list items to show advertisement Sitelinks.

4. For Location Extensions that are not appearing, ensure that the area you've entered in the Ad Preview Tool is inside your focused-on area.

6.34 Best Practices Like Using Features Such as Reviews, +1 Button, etc.

The most ideal approach to persuade somebody to purchase your item/administration is to make your other client's opinion on your item/administration accessible to them. Many times, when you buy something on the web, you check for the client reviews and judge the item. Consequently, we can say that online surveys have risen as one of the main considerations that impact the buying practices of the present buyers. Surveys convey intense verbal power that can draw or drive away potential visitors. Interpersonal organizations like Facebook, Google+, and Foursquare enable visitors to rate and survey nearby organizations. Organizations presently hold onto surveys as a major aspect of their promoting technique, and conducted appropriately, online surveys can likewise support and improve the viability of your web-based social media strategies [28–32]. To make the most of this strategy, the following points can be taken into consideration:

a. **Identify most relevant online review sites**: Every business can have its own type of arrangement of web-based life survey sites. That is the reason it's critical to realize who is discussing your image and where. Web-based life is earned media, and achievement requires tuning in. Be that as it may, to have the option to tune in to your visitors and your intended interest group, you initially need to know where their discussions are occurring. This is the reason it's so basic to tune in on all the conceivable social sites, channels, and stages that are significant to your business. You may as of now have a business page on Facebook and a marked profile on Twitter; however don't stop there. Plant your banner on survey sites and other social applications too. Yelp, Google+, TripAdvisor, Foursquare, Citysearch, and Yahoo Local are only a couple of the spots your business could be referenced. When you've done that, distinguish and guarantee your business on specialty explicit sites where more visitors might talk. For instance, Zomato, OpenTable, and Zagat are perfect for eateries, while Cars.com and DealerRater are for car organizations. In case you're a medical supplier, Vitals and Healthgrades are basic survey sites to know. This strategy sounds straightforward enough; however few use it.

b. **React to your online reviews**: There is no reason to fear the voice of the client. While negative online networking remarks and one-star reviews feel like a punch in the gut, you need to confront client criticism head on. The savviest brands know that, to prevail via web-based networking media, you need to create significant associations with your fans and adherents. This even incorporates your greatest pundits and haters. Reacting enables you to engage with the vocal client who composed the review. It additionally creates an open door for unprejudiced onlookers, your potential visitors, to have a more pleasant, progressively reasonable impression of your brand image [29].

c. **Urge happy customers to give positive reviews**: An incredible method to get your positive reviews moving is to have your upbeat visitors start the discussion. At that point you can share your best reviews via web-based networking media systems. The idea that your most joyful visitors may be withdrawn, that their quiet could be affecting your image notoriety, is disappointing. There's additionally no requirement for it to work out as expected. Via web-based networking media, where user-generated content (UGC) rules, you can shape your system in a manner that motivates trust and empower your community to speak. Rather than concentrating on special brand content or paid promoting as your top venture needs, you can let UGC do something amazing for your image notoriety. An extraordinary strategy for utilizing UGC is to gather criticism from your most joyful visitors and urge them to post audits. These individuals can be incredible informal impetuses for your business, so it bodes well to keep them drawn in [22].

d. **Advance your business on review and social media sites**: By its tendency, online networking makes chances to exponentially extend your system, broaden your brand image's reach, and develop your group of spectators. Try not to restrain your endeavors to "Like us on Facebook" or "Follow us on Twitter". Let individuals realize they can likewise discover you on Yelp, TripAdvisor, Google+, Foursquare, and different stages. This acts as an exhibition of your authenticity, straight forwardness, and receptiveness to input, which can be the most basic factor to prevailing via web-based networking media [24].

e. **Recognize and resolve customer issues**: In the event that your visitors have raised issues via web-based networking media in online reviews, you have to find these issues and resolve them. Web-based life has changed how organizations handle grievances and react to visitors' needs and desires. Outside of Facebook and Twitter, the voice of the client can likewise be heard on network-based, survey-driven stages. In the event that you aren't as of now giving "social" care and backing, this point proves to be an interesting [21].

f. **Influence feedback and reviews to gain customer insights**: Much the same as UGC via web-based networking media, client input and online surveys can be tackled to enable your business to find and decipher key bits of knowledge on what and how visitors think. Bolster the sites of your web-based social networking analytics by taking advantage of your valuable survey information.

g. **Offer positive feedback with your team**: Behind probably the savviest brands via web-based networking media are associations that promise the client an experience. If your social media strategy effectively drives client commitment, creates five-star surveys, and your profiles thrive with hearts, preferences, and top picks, spare a moment to share and celebrate with your group. In addition to the fact that this boosts staff confidence and representative fulfillment, it can likewise encourage energetic and ground-breaking brand evangelists in your group. Engaged, happy, and persuaded workers work well as the voice of your image in web-based social media interactions, while making your brand image irresistible to your community.

6.35 Offer Management Plan

6.35.1 First Let Us Comprehend What Is PPC Offering

The dollar sums an organization designates to pay-per-click publicizing are known as pay-per-click (PPC) offers. Offer costs change contingent upon the organization's publicizing the spending plan and the time the organization intends to utilize paid inclusions. Other integral elements are the specific keyword prominence and rivalry from different promoters. The more prominent the focused-on specific keywords, the higher the PPC offer should be to verify the space on the web search tool pages. PPC offers pay for the ads that numerous individuals see on an internet search engine results page, for the most part at the top or as an afterthought, and are independent from normal outcomes. Common outcomes are sites that position through website improvement strategies. The advertisements that PPC offers pay for are for the most part called supported connections or supported promotions, and show up when the specific keyword that the PPC offer obtained is gone into the pursuit question. There are likewise programming projects for the executives of PPC ventures at Google AdWords, Yahoo! Search Marketing, and MSN AdCenter in one simple interface. The management of PPC campaigns is tedious and requires close checking to guarantee that the financial backing designated is legitimized. The software management programs claim that their advanced bidding algorithm saves costs and most have a comprehensive reporting function that monitor the return on the investment.

We have to remember the following simple points before continuing ahead:

1. Cost-per-click (CPC) offering is the point at which you set the most extreme cost you're willing to pay for somebody tapping on promotions. This offering strategy expects you to pay just when a watcher is intrigued enough to click your promotion and find out additional information. This is extraordinary for direct reaction objectives.

2. Cost-per-thousand impressions (CPM) involves paying for each arrangement of a thousand showings of your advertisement as opposed to paying for the quantity of clicks that you get. CPM bidding is best for brand awareness as opposed to direct response or directing people to your site.

3. A cost-per-acquisition (CPA) offering, accessible through AdWords Conversion Optimizer, enables you to set the amount you need to pay per transformation by determining either a maximum CPA or a target CPA for every promotion gathering. Conversion Optimizer finds you the most potential changes for your financial limit inside your maximum or target CPA. This is an incredible decision if your objective is transformations.

4. Enhanced expense per click (ECPC) is an offering highlight that raises your offer for clicks that appear to be bound to prompt a change [31]. Presently you may have had a reasonable idea of how to utilize an offering deliberately; let us jump into it further for a better understanding of the offering system.

6.35.2 Comprehend Offering System

Understanding conversion systems: As observed, the idea is to deftly mechanize the precise offers that your campaigns place in the auctions, in light of how likely a specific keyword is to prompt a transformation.

6.35.3 Revenue/Conversion Strategies

ROAS = (revenue)/(cost): Most PPC management platforms will work to restore a set ROAS for the sum spent on your campaigns. Individual conversion values can shift, yet your device will average out the expenses to meet the ROAS you set. It is imperative to screen your ventures utilizing ROAS procedures to ensure you aren't smothering regions where you can stand to spend more. Advertisers regularly find that conversions with low ROAS drag pulling down conversions with high ROAS when they are packaged in one campaign. It's a smart thought to bunch campaigns or promotions with comparative ROAS objectives. This kind of bidding system is incredible for organizations with producer services that are worth varying amounts. If you sell vehicle parts, for instance, actualizing an ROAS methodology will average the income you create over the majority of your items. Gather together items that ought to create comparative measures of ROAS and let your PPC device take every necessary step.

6.35.4 Perceivability Strategies

It's not constantly about change objectives. Now and then expanding perceivability in key regions is progressively significant.

6.35.5 Search Page Location Strategy

This sort of system enables you to set and overlook your inquiry offers. Your administration stage will change offers for the duration of the day until your spending limit is reached. While this methodology can be helpful, it can without much of a stretch drive up your expenses while gutting the benefit of your ventures. Search sell-offs are consistently in transition; the expenses per click change drastically based on things like device, time of day, and sale members.

You should implement a target search page location strategy when you are open to burning through cash that probably won't return direct income. Otherwise, you should keep an exceptionally close eye on your campaigns, and should change your technique to something less unpleasant. Search page area methodologies are extraordinary when you have cash to spend. The problem is the shrinking or once in a while non-existent productivity that you may see. That is the reason this procedure is helpful when your essential objective is perceivability and acknowledgment. You will spare yourself a migraine by holding these sorts of campaigns to a higher standard in comparison to the rest of your record. Concentrate on measurements like impression share and assisted conversions rather than direct transformations and ROI. The objective is to get your name out there and impact changes not far off, so take a step back and think longer term.

6.35.6 Outranking Share Strategies

These are utilized when you need to outrank a contender and catch more taps by and large. Actualize them to ensure that your advertisements as often as possible show more than those of a rival in indexed lists. Outranking share techniques can be unfathomably compelling when folded into a bigger competitor strategy. Positioning is the name of the game here. Beside the upside of a higher situation in the advertisement sell off, the immense incentive in this methodology is the capacity to connect your image with that of your rivals. Your potential guests see your advertisement over your rivals (which builds trust) at whatever point an inquiry is made. This is a chance to deliberately publicize the contrasts between your brands. Numerous little brands competing with enormous players utilize this methodology with extraordinary outcomes. The key is to compose an outstanding promotion duplicate. Outranking an adversary isn't worth very much if their promotions are superior to yours. Branding and its associated activities are indistinct naturally, so it tends to be troublesome when estimating your success.

6.35.7 Click-Based Strategies

Maybe the most immediate of all, this procedure will amplify the volume of clicks accessible for your financial limit. You won't have the option to set any CPA or ROAS targets, however. You should watch out for your ventures to ensure they are beneficial. While amplifying clicks is less advanced than different procedures examined above, there is as yet potential for an advertiser who has profound information of their record.

Selection of click-based strategies is done on the basis of the following two segments:

1. **Financial limit or Budget**: How much would you say you will spend on the campaign ? Keep in mind that you pay for each promotion click, so you are basically advising your tool to charge you as much as would be prudent.
2. **Max CPC**: How much would you say you will pay per click? Your maximum CPC will decide the measure of clicks that you will get. A tick procedure is easy to execute, which makes it amazingly flexible. It is simpler to expound on circumstances when it would not work for you. For instance, accounts with thin margins would not

be great up-and-comers. You chance rapidly overspending in territories where you are not benefited, particularly in the event that you
are bidding on important terms or other costly specific keywords.
Ensure you keep your CPCs under control. Volume is the name of
the game, so advertisers pushing eyeballs to sites that profit on promotion impressions have utilized this technique with progress.

6.36 Manual vs. Computerized Bid Management

Regardless of whether you utilize a robotized or manual framework, it truly
takes a decent parity of human and PC association to create the best outcomes. Continuously set aside some effort to examine your outcomes and
make course revisions when important. Try not to feel that you can simply
kick back and appreciate the ride in the event that you utilize a computerized
framework.

6.36.1 Manual Offering

Manual offering includes dealing with your offers directly through AdWords
or Bing, which means making offer increases or decreases dependent on a
few factors, for example, past keyword execution or advertisement position,
and not depending on mechanized arrangements. It is a manual procedure
that requires human instinct, and it very well may be contrasted with being
in the everyday channels where you are totally associated with each change
and subtlety of your record. The sort of advertisers who lean toward this
kind of offering esteem the most elevated level of control and the capacity to
roll out quick improvements, be they small or enormous.

6.36.1.1 Focal Points of Manual Offering

1. **Instant responses**: Suppose while observing your record, you see
   that performance of campaign response has all of a sudden dropped
   today. You run an Auction Insights report and presume that new
   contenders have been pushing CPCs to increase, so you rapidly push
   offer changes for the specific keywords that have been performing inadequately to counter the current situations. The capacity to
   respond to the continually changing conditions on the web immediately is an immensely favorable position.

2. **Control**: With this sort of offering, you can offer on the individual
   keyword level and decide how forceful you need to be with your
   offers at the most explicit level. It is power and granularity at its best.

3. **No postponement in changes**: Manual offering guarantees that your progressions produce a result immediately. The fact of the matter is that you likely need more information for a mechanized framework to have a critical effect. Mechanized apparatuses improve when there's a bigger dataset to work with.

6.36.1.2 Impediments of Manual Offering

1. **The bigger the record, the bigger the test**: Huge spending accounts that house a huge number of keywords (particularly those in E-business) become increasingly harder to successfully oversee on a granular level.

2. **Chance of wasteful aspects**: Having an everyday perspective on the record can conceivably keep an advertiser from seeing the "greater" picture. Likewise, there is the danger of human error due to the human segment inside manual offering.

3. **Bidding can gobble up your time**: For PPC supervisors who don't utilize offer administration programming, altering offers physically can without much of a stretch transform into an all-day job, so the inquiry lies in gauging the opportunity costs and advantages for both offering choices and making sense of which suits you and your records best.

4. **Limited division choices**: If you need to offer certain ventures or promotion bunches comparatively, you are constrained to a specific record structure. With robotized offering, you have the adaptability to structure your record any way and mix various campaigns and promotion bunches in any blend you want for offering purposes.

6.36.2 Computerized Bidding

An offer administration instrument enables you to oversee PPC ventures over various web crawlers and gives a central source for tracking and analysis. Sooner or later it turns out to be too hard to even think about managing the majority of the related complexities. There is a solid discussion out there on whether offer administration apparatuses are extremely powerful. It comes down to the multifaceted nature of your campaigns, and in the case of utilizing, an offer administration device will enable you to set aside time and cash. Commonly the cost, execution, and multifaceted nature of the campaign are too high to even think about justifying, given the outcomes. Essentially you have to represent the time and cost related with dealing with your venture, and think about how the different devices available will influence both of those. By and large, in the event that you feel extended you should give it a shot. Most device merchants offer a free trial, so you can figure out its adequacy before you purchase. Offer administration can be a helpful instrument, yet it is imperative to bear in mind that in the present promoting condition, you have to do significantly more, notwithstanding what the offer administration device can do. For example, advertisement duplicate and greeting page testing and tuning won't profit by an offer administration instrument. You have to have different procedures set up to deal with these. Then again, utilizing an offer administration device to deal with your long-tail keywords can pay enormous profits. Long-tail specific keywords are less aggressive than the significant ones, yet expect you to actualize colossal quantities of keywords.

6.36.2.1 Apparatuses Accessible for Computerized Offering

1. **Google Conversion Optimizer**: If you're searching for an apparatus to enable you to deal with your offers and you are just utilizing Google, at that point you should consider the Google Conversion Optimizer. Similarly, as with a large portion of Google's devices, it's free in the event that you are utilizing AdWords. Be that as it may, you have to accomplish 30 changes in the previous 30 days to utilize it.

2. **PPC Bid Max**: Formerly Dynamic Bid Maximizer, PPC Bid Max is an offer administration apparatus intended to enable you to oversee campaigns with the three big web search tools – Google, Yahoo!, and Bing.

3. **Keyword Max**: Another device for dealing with your offers with the majority of the significant web crawlers inside one interface is Keyword Max. Notwithstanding its offer administration instruments, Keyword Max likewise offers a large group of other PPC apparatuses like specific keyword age devices.

4. **Omniture**: Omniture has built up an extraordinary notoriety as an SEM device seller. They too have bid management tools, just as a large group of others that will enable you to deal with your PPC campaigns. In spite of the fact that they might be somewhat expensive, the suite of devices they offer will assist you with numerous different parts of your ventures.

5. **Other devices**: Clickable, Marin Software, Search Force, Kenshoo, and Acquisto.

6.36.2.2 Focal Points of Automated Offering

1. **Versatile division**: The underrated capacity to section your campaigns and advertisement gatherings dependent on likeness and classify them into custom folders which you would then be able to offer independently.

2. **Efficiency**: Robotization can give you the opportunity in data transfer capacity to deal with your enormous records on a full-scale level and spotlight rather on development openings or record system while likewise monitoring ordinary execution.

3. **Ability to deal with huge or complex records**: It tends to be a challenge to deal with the majority of the great many specific keywords inside huge or complex records, so mechanization can help in productivity gains.

6.36.2.3 Inconveniences of Automated Offering

1. **Traffic and transformation recommendations**: It is exhorted that exchanging over to mechanized offering is best when your record has average traffic and change volume ...but not before.

2. **Automation doesn't mean hands off**: By utilizing these platforms(s), despite everything, you ought to effectively watch your record and pull levers all the time to ensure these instruments are adequately playing out the manner in which you expect them to. You additionally don't need the instrument to carry out your responsibilities for you, but rather supplement your current pursuit activities.

3. **Potential deferral in changes**: Since you make changes in the stage and not legitimately in the UI, there is an extra step of syncing your changes to the channel. Most stages will naturally synchronize once or a couple of times each day, with the alternative to physically adjust your campaigns, so if you neglect to match up your progressions quickly, they won't be in actuality until the stage's next booked match up.

4. **Be careful about forceful changes**: When upgrading with an offer administration instrument, it is savvy not to be excessively forceful when altering targets because algorithms require time to create and fabricate history to decide the ideal offering so unexpected huge changes may contrarily influence execution.

6.37 Diverse Bid Management Features Like CPA Bidding, Position Preference, etc.

6.37.1 Cost-Per-Acquisition Strategy

Utilize a CPA technique when you have explicit objectives that you need your site guests to accomplish. Setting up transformation focuses in your site is clear; however doing so can require significant investment and a great deal of research. You need your transformations to be as immediate and simple to accomplish as can be expected under the circumstances. Adding things to shopping baskets or pamphlet structure entries are great possibilities for a CPA advancement methodology. The best ones are consistently yes or no parallel activities – the guest hit the catch, or they didn't.

6.37.2 Upgraded Cost-Per-Click Strategy

You should execute an eCPC system when you aren't open to giving up the majority of your control to a PPC apparatus. This methodology is significantly less uninvolved than the CPA-style offering system, yet despite everything it offers a ton of robotization. Your campaign will do all that it can to build transformations, yet it will take different restrictions like max offers when doing as such.

6.37.3 Position Inclination

With the Position Preference included, you could choose a particular situation as an objective. In the event that you had verified that you accomplish most elevated ROI with positions 3 through 4, for instance, you could set your position inclination to these positions. The framework at that point attempts to demonstrate your advertisement in positions 3 through 4 by expanding or bringing down its offer. Testing your advertisements at different normal positions from the get-go in your venture's lifetime makes you the best judge of which position gets you the greatest changes at the best CPA and most extreme ROI. When this ideal normal position has been determined with your endeavors, the Position Preference highlight is utilized to keep up your promotions at those positions. This would enable you to get the greatest number of transformations at the most rewarding ROI [21].

6.38 Conclusion

It's imperative to see how algorithms function to apply context to what you're experiencing/reading. Knowing this helps with deciphering which parts of a site or the world are being balanced in an update and how that alteration fits into the enormous goal of the engine. Further, moving forward it's critical to understand entities: Knowing this will enable you to comprehend not only exactly what content is significant (how close are those elements you're expounding on?) yet additionally which connections are probably going to be judged more positively. What's more, that is simply to name a few points of interest. Search calculations work as a huge gathering of different calculations and recipes, each with its very own motivation and goal, to create results a client will be happy with. Truth be told, there are calculations set up to screen only this part of the outcomes and make alterations where positioning pages are thought not to fulfill client expectations dependent on how clients interface with it. Included in this are algorithms designed specifically to understand entities and how entities relate to each other in order to provide relevancy and context to the other algorithms. The site designer ought to know about different SEO methods that step by step guarantee that the profoundly reliable content-rich site is set in the top-positioned result pages of the internet search engine. The correlation investigation of SEM and SEO helps us to comprehend the promoting and specialized part of the site in advanced advertising. The keyword investigation plays an indispensable role in the predominant SEO procedures. The web-crawled texts and connections, indexed pages, and the keyword density of the site can be observed utilizing progressed SEO apparatuses and give the SEO report on a day by day or week by week or month to month basis, which improves the traffic of the site and the general offers of the item or administrations.

References

1. Asdemir, K.U.R.S.A.D., & Yahya, M.A. (2006). Legal and strategic perspectives on click measurement. *SEMPO Institute Opinions and Editorials*, 11, 1–11.
2. Jansen, B.J., & Resnick, M. (2005, June). Examining searcher perceptions of and interactions with sponsored results. In *Workshop on sponsored search auctions*.
3. Barry, C., & Charleton, D. (2008, July). In search of search engine marketing strategy amongst SME's in Ireland. In *International conference on E-Business and telecommunications* (pp. 113–124). Berlin: Springer.
4. Zhang, L., Zhang, J., & Ju, Y. (2011, May). The research on search engine optimization based on six sigma management. In *2011 international conference on E-Business and E-Government (ICEE)* (pp. 1–4). IEEE.
5. Terrance, A.R. (2017, February). Search engine optimization–a critical element in digital marketing. In *National seminar proceedings of a paradigm shift towards empowerment of women* (pp. 66–72).
6. Fain, D.C., & Pedersen, J.O. (2006). Sponsored search: A brief history. *Bulletin of the American Society for Information Science and technology*, 32(2), 12–13.
7. Feng, J., Bhargava, H.K., & Pennock, D.M. (2007). Implementing sponsored search in web search engines: Computational evaluation of alternative mechanisms. *INFORMS Journal on Computing*, 19(1), 137–148.
8. Jansen, B.J., & Molina, P.R. (2006). The effectiveness of Web search engines for retrieving relevant ecommerce links. *Information Processing & Management*, 42(4), 1075–1098.
9. Subhankar, D., & Anand, N. (2019, May). Digital sustainability in social media innovation: A microscopic analysis of Instagram advertising & its demographic reflection for buying activity with R. In *1st International scientific conference "Modern Management Trends and the Digital Economy: from Regional Development to Global Economic Growth" (MTDE 2019)*. Atlantis Press.

10. Singh, I., Nayyar, A., Le, D.H., & Das, S. (2019). A conceptual analysis of internet banking users' perceptions. An Indian perceptive. *Revista ESPACIOS, 40*(14), 1–17.
11. Mohanty, P.C.; Dash, M.; Dash, M., & Das, S. (2019). A study on factors influencing training effectiveness. *Revista Espacios, 40,* 7–15. Retrieved from http://www.revistaespacios.com/a19v40n02/19400207.html
12. Singh, I., Nayyar, A., & Das, S. (2019). A study of antecedents of customer loyalty in banking & insurance sector and their impact on business performance. *Revista ESPACIOS, 40*(06), 11–28.
13. Singh, S., & Das, S. (2018). Impact of post-merger and acquisition activities on the financial performance of banks: A study of Indian private sector and public sector banks. *Revista Espacios Magazine, 39*(26), 25.
14. Das, S., Mondal, S.R., Sahoo, K.K., Nayyar, A, & Musunuru, K. (2018). Study on impact of socioeconomic make up of Facebook users on purchasing behavior. *Revista Espacios, 39,* 28–42. Retrieved from http://www.revistaespacios.com/a18v39n33/18393328.html
15. Mondal, S., Das, S., Musunuru, K., & Dash, M. (2017a). Study on the factors affecting customer purchase activity in retail stores by confirmatory factor analysis. *Revista ESPACIOS, 38*(61), 30–55.
16. Mondal, S., Mall, M., Mishra, U.S., & Sahoo, K. (2017b). Investigating the factors affecting customer purchase activity in retail stores. *Revista Espacios, 38*(57), 22–44.
17. Mentz, G.S., Mentz, G.S., & Whiteside, R. (2003). The revenue impact of online search engine marketing technology for medical clinics. *Journal of Information Technology Impact, 3*(2), 101–110.
18. Singh, S., Mondal, S., Singh, L.B., Sahoo, K.K., & Das, S. (2020). An empirical evidence study of consumer perception and socioeconomic profiles for digital stores in Vietnam. *Sustainability, 12*(5), 1716.
19. Singh, L.B., Mondal, S.R., & Das, S. (2020). Human resource practices & their observed significance for Indian SMEs. *Revista ESPACIOS, 41*(7). Retrieved from http://www.revistaespacios.com/a20v41n07/20410715.html
20. Sharma, E., & Das, S. (2020). Measuring impact of Indian ports on environment and effectiveness of remedial measures towards environmental pollution. *International Journal of Environment and Waste Management, 25*(3), 356–380. doi:10.1504/IJEWM.2019.10021787
21. Das, S. (2020). Innovations in digital banking service brand equity and millennial consumerism. In *Digital transformation and innovative services for business and learning* (pp. 62–79). IGI Global.
22. Mondal, S.R. (2020). A systematic study for digital innovation in management education: An integrated approach towards problem-based learning in Vietnam. In *Digital innovations for customer engagement, management, and organizational improvement* (pp. 104–120). IGI Global.
23. Nadanyiova, M., & Das, S. (2020). Millennials as a target segment of socially responsible communication within the business strategy. *Littera Scripta, 13*(1), 119–134. doi: 10.36708/Littera_Scripta2020/1/8
24. Mondal, S., & Sahoo, K.K. (2020). A study of green building prospects on sustainable management decision making. In *Green building management and smart automation* (pp. 220–234). IGI Global.

25. Das, S., & Nayyar, A. (2020). Effect of consumer green behavior perspective on green unwavering across various retail configurations. In *Green marketing as a positive driver toward business sustainability* (pp. 96–124). IGI Global.

26. Das, S., Nayyar, A., & Singh, I. (2019). An assessment of forerunners for customer loyalty in the selected financial sector by SEM approach toward their effect on business. *Data Technologies and Applications, 53*(4), 546–561. doi: 10.1108/DTA-04-2019-0059

27. Jones, W.D. (2006). Microsoft and Google vie for virtual world domination. *IEEE Spectrum, 43*(7), 16–18.

28. Omprakash, K.S. (2011). Concept of search engine optimization in web search engine. *International Journal of Advanced Engineering Research and Studies, 1*(1), 235–237.

29. Kumar, V., & Shah, D. (2004). Pushing and pulling on the internet. *Marketing Research, 16*(1), 28–33.

30. Laffey, D. (2007). Paid search: The innovation that changed the Web. *Business Horizons, 50*(3), 211–218.

31. Lempel, R., & Moran, S. (2000). The stochastic approach for link-structure analysis (SALSA) and the TKC effect. *Computer Networks, 33*(1-6), 387–401.

32. Bansal, M., & Sharma, D. (2015). Improving webpage visibility in search engines by enhancing keyword density using improved on-page optimization technique. *International Journal of Computer Science and Information Technologies, 6*(6), 5347–5352.

7

Multivariate Testing Remarketing and AdWords

7.1 MVT and Advertisement

This MVT test is a tool to a hypothesis in which multiple variables are modified. It is a specific bi-variate test that includes the coherent perceptual ability of the viewer and analysis of more than one outcome variable in consideration [1]. So basically, here calculations are not bi-variable, they are based on multiple variables. The bi-variate test used to measure the impact of progressively focused changes the while multivariate test helps to quantify the chronological impact due to dynamic changes [2]. There are three types of MVT: Multivariate, multivariant, and multivariable testing. Here the onsite webpage and its elements are considered only.

The point of view of the developer for the website is the main aspect of consideration. The feature on the page can basically change the rates of AdWords on the landing page. One can go for a few variations of words and test effectively to check what works and what doesn't. For example, a banking website can promote with taglines "Debit Relief That Works", "Free Yourself from the Burden of Debit", or "Get Relief from Debit". One has to see what best suits.

Multivariation is like one variable with numerous modifications or variations [3]. Multivariate analysis tests the changes one item or one variable but gives many modifications or variations. Example: A bank wants to promote *"Debit relief that works with a lady in picture" or changes it to "Get relief from Debit along with a male employee's picture"*. Here one can change the caption and picture as required. So, two variations of one caption and image can be used.

In a multivariate test, we control the change with two variables with their respective changes in content [4]. The image and caption both can be changed with their variation. Example: A bank wants to advertise "Debit relief that works" and "Get Relief from Debt" both with an image of a woman and an image of a man. So, here there are two variables giving four options.

So MVT tests various variations of variables to get the optimized page for a site. So, it depends on the developer to get the best picture and image which can get maximum traffic on the landing page. So, to control the tool, we have to look for the promotional expenditure plan for the present web page which is targeted. So, the developer can go with the control of MVT by putting two extra variations along with the primary one for the client [1,4].

Different duplicate MVT tests two different things for a promotional campaign such as traffic and change of content. Sometimes it becomes inappropriate too in these five cases:

1. When time and traffic are limited or not available.
2. Weird pattern of visitors.
3. Used up important assets.
4. No learning pattern in web series.
5. When MVT can't give a framework.

MVT takes a lot of time to judge the traffic and change. Low traffic can't give a good performance, so if it is taken then, the results will be ambiguous. So, significant measurement will come out only after prolonged exposure and more interaction with traffic. Factual results depend on quantitative analysis. MX Toolbox with IT enabled services and DNS server help in the analysis of MVT as they help quantitative analysis on websites with a lot of traffic and sharing [2–3].

MVT goes for eight varieties for three variables, then two security pictures in Captcha, then traffic for each variable will be decreased 6.25%. As we saw, only 3 factors with 2 variations brought about 8 varieties, and adding 2 greater security trust images to the blend carried this to 16 mixes. Traffic to every variety would be diminished to only 6.25% [1].

7.2 Significant Test of More Varieties to Identify Larger Errors in Promotional Campaign

As per the conventional guidelines, the conversion testing technique permits close to six varieties for any A/B test (bi-variate test or split test) on the grounds that the safety buffer turns into an issue. In an A/B test with 2 varieties, we might have the option to arrive at factual criticalness in about 14 days, and bank a 10% expansion in transformations. In any case, in a test with six varieties, we may need to keep running for about a month before we can accept that the 10% increase is genuine. The room for give and take is bigger with six variations requiring more time to reach statistical significance.

Presently consider a multivariate test with many varieties. Larger and larger margins of error mean the need for even more traffic and some special calculations to ensure we can believe our results aren't just random. Eventually, the majority of these varieties do not merit testing. Think about a multivariate test as a framework that consequently makes every conceivable blend to enable you to locate the best result. So superficially, it sounds engaging [5].

So, MVT has two potential probable outcomes:

1. Loss of client can happen over variation varieties.
2. Time and energy should be used more to do the analysis between all varieties and variations.

Missing the development of different variations and varieties becomes very expensive as one never knows which one will be suitable. We can control the cost in two ways, preferential control of variations and guiding traffic to different varieties, which means the test will require less effort to arrive at essentialness [6]. At the point when tests run quicker, we can test all the more every now and again. Then again, multivariate tests go through all varieties, or an enormous sample of varieties.

Actually, it is very difficult to learn from multivariate tests, as when they are conducted the content is already in the public domain and it takes a huge effort to get the views calibrated [5]. If MVT finds success, would we be able to know why? What would we be able to reason from this? Which component was most essential to our guests? For example, the addition of trust symbols and subsequent effect. Furthermore, for what reason does it make a difference? First off, it makes it simpler to think of good test speculations later on. If we realized that including trust symbol had the greatest impact, we may choose the addition of a trustworthy image and content. Lamentably, we can't rely always on the not so clear, conspicuous idea. When you take in something from a trial, you can apply that idea to different components of your site. On the off chance that we realize that the arrival strategy was a main consideration, we may take a stab at including the arrival arrangement for webpages. The arrival to the landing page can also be added. Testing isn't just about discovering more income; it is also about the understanding of visitors' preference [3,5].

MVT is very enticing. It often attracts the mindset of testing a lot of variations along with content, but it is always not possible. If a lot of testing happens, then content may lose the originality and meaning of representation. Rather these tests could be targeted to find the exact reason for getting clicks, generate practical compatibilities with content, assume and design research for knowing the exact reason of assumption, analyze the pattern of traffic, calibrate the results, and go for futuristic scope of developments. A few obstacles, like wrong reason identifications, conspicuous speculations, and dubious content, may create problems for MVT [7].

7.3 Artificial Intelligence and MVT

From recent developments of artificial intelligence (AI), it is evident that AI will give a great helping hand to MVT. Neural systems along with the synergistic action of computers prove useful in giving a great boost to MVT. This integration of AI is called a transformative neural network (TNN) or hereditary neural network (HNN). It helps in giving the results in less time by studying the probability of possible potential variations [4,7]. All the developmental calculations sort through possible variations, selecting what to test so that we don't have to test all combinations. These evolutionary algorithms follow branches of patterns through the fabric of possible variations, learning which are most likely to lead to the highest converting combination. Poorly performing branches are deleted for almost certain winners. Over time, the highest performer emerges and can be captured as the new control. These calculations additionally present transformations [4]. This technic gives guaranteed outputs quicker than the physical traffic. Transformative neural systems enable testing instruments to find what mixes will work without testing every single multivariate mix. With AI, MVT is very much futuristic.

7.4 Remarketing Fundamentals

In 2010, remarketing was implemented into AdWords. From that point forward, it has turned into a solid staple in a huge number of insightful businesspeople's showcasing collections [8]. With dynamic remarketing, you can advance your advertising effort for the greatest possible number of transformations [5,8]. With dynamic remarketing, you can basically indicate explicit promotions to guests who have just visited your site. Being able to reach your target market as deeply down the funnel as possible can ensure that each of your ads is as relevant as possible, can guarantee that every one of your advertisements is as applicable as would be prudent [6–7]. Remarketing gives precision and clarity to advertisement in for both static image and dynamic video-based adverts. The only thing that is in any way important is that in all likelihood your clients are set apart with a unique code which helps in remarketing.

7.4.1 The Most Effective Method to Set up Unique Remarketing in AdWords

Each and every day, a large number of individuals start up E-business sites. Over the whole globe, the most widely recognized type of E-trade is business to customer (B2C), which is also referred to as online shopping or retail, or in short E-commerce or E-retail or E-tail activities, irrespective of whether the business fully or partially operates online. Here dynamic remarketing is required to enhance the advertisement. Despite having some minute problems with operation, utilizing dynamic remarketing is justified as the web-based shopping business sector is very solid and gives no indications of fading away at any point in the near future. It is estimated to be at $224 million by end of 2019 [6].

New E-com websites find it very hard to penetrate the already crowded market. So, with dynamic remarketing they can get the best out of the E-commerce market place as it allows adverts to be streamed to those who have visited the website or use the app on mobile [7]. Dynamic remarketing lets you show previous visitors and customers ads that contain products and services they viewed on your site. After careful analysis of the data, you can go for dynamic remarketing where the main aim objective will be which visitors will be targeted for the adverts. A new E-com website can segregate the visitors in three categories, for example, those who are avoiding certain webpages, those who frequently visited webpages, and specific product pages which they visit repeatedly [7,9]. Other than this, customized visitor segregation can be done for specific websites, for example who has kept which item in their cart before checking out in the near future [10]. Other elements of considerations apart from dividing the customers will be, for example, potential future parameters like time of visit, best possible URLs, and analysis of the entire sales funnel from different touch points.

7.4.2 Purpose of Remarketing

With the help of Google Analytics, one can go on setting the remarketing objectives such as:

1. To measure webpages that visitors have visited.
2. To find the time that visitors spend on the website.

7.4.3 Setting up of Dynamic Remarketing with AdWords

1. Firstly, find who has clicked on campaigns run on websites.
2. Secondly, click on the **+Campaigns** button and create a new campaign. **Choose Display Network only.**
3. Thirdly, on the next screen, you check the **Buy on your website** box inside the **Marketing objectives.** Now choose the location. After this, you need to **define your bidding strategy and choose your budget.**
4. Name the remarketing ad group and set the max CPC bid.
5. Tap on interest and remarketing, then on the drop-down menu, pick remarketing records and click **lists** and click on **Use Dynamic**

remarketing ads. Of course, you select **Retail** in the other drop-down menu below. Either the developer creates remarketing or AdWords will create a tag for the website when configured by its own tag [10]. AdWords will make four to five default remarketing options for the website.

6. To complete the manual setup, select the correct feed for remarketing ads tag and select the appropriate one.

7. After clicking on save and proceed at the bottom, you will be able to remarket your adverts. You can choose every single accessible design, configuration, and size.

8. In the wake of making your advertisements, there is just one stage to go: Just click Save and keep on promoting! No stress, you'll see a total audit of your promotion set before continuing. In the wake of choosing the billing address and subtleties, your advertisements are prepared to go!

7.4.4 Dangers of Remarketing

Taking everything into account, we can say that the potential advantages of remarketing effectively are huge. Likewise, with some other promoting system, there is an undeniable component of hazard [10]. At times, you may need to increase your offer marginally to guarantee that your underlying advertisements can have as much exposure on publishing sites as you want. It's always astute to do a starter low-speculation test to check whether you can measure any huge distinction between an ordinary promotion campaign and a remarketing effort [11].

7.5 Most Common AdWords Display Ad Sizes

The common AdWords display ad sizes are as follows:

1. 250 × 250: Square
2. 200 × 200: Small square
3. 468 × 60: Banner
4. 728 × 90: Leaderboard
5. 300 × 250: Inline rectangle
6. 336 × 280: Large rectangle
7. 120 × 600: Skyscraper
8. 160 × 600: Wide skyscraper

9. 300 × 600: Half-page ad
10. 970 × 90: Large leaderboard

The important ad size that cannot be accessed on desktop PC work area or tablet and can be viewed at the top of mobile screens is 320 × 50 – mobile leaderboard.

The top performing AdWords banner sizes are as follows:

1. 300 × 250: Medium rectangle
2. 336 × 280: Large rectangle
3. 728 × 90: Leaderboard
4. 300 × 600: Half page
5. 320 × 100: Large mobile banner

Max File Size:

Since files more than 150 kb in size take more time to stack, Google won't permit files larger than this for advertisements. The advert files should be in .jpeg, .swf, .png, or .gif since Google doesn't consider other file formats.

7.5.1 Need for Display Advertising

It is not that the adverts are always able to highlight the inner message of content. Display adverts normally get lower direct click through rates than search ads raising brand awareness and remarketing for previous visitors to raise the likelihood of generating clicks. They normally *push* the visitors into business channels on the principle of attention, interest, desire, and action (AIDA) [12]. The working principles and guidelines are as follows:

1. Rules for document size of max 150 kb with fringe diagrammatic view so that adverts can be isolated from the base of website they work on.
2. Adverts should have less than or equal to 20% of content including logos to attract the traffic and go with the promotional campaign as that happens in social media like Facebook.
3. The content of adverts should be understandable and reciprocate with the audience.
4. The use of *hues* to make the content attractive with no animation so that they will lead the visitor to the core main theme of the product and not to the outliers and peripherals [2].
5. Adverts should have a *clear* inspiration so that CTA looks like a button that stands out. For example, the developer can use "Shop Now >>" or "Find out More!"

6. The image and design will make the webpage nice to look at and attractive.

7. Use of *symbolism* to ensure visitors pick up your unspoken unwritten message.

8. The content and adverts should be clear and precise. Excessive use of anything can ruin the campaign.

9. Before uploading, the developer should test the advert or simulate it so that sustainability and viability will increase.

7.5.2 Rich Media Formats for Various Networks across the Desktop and Mobile Platforms

1. **Intelligent adverts in banners**: The content should speak about the position and intelligence of content that can attract interactive traffic for the product. They can be attractive to the visitor.

2. **Billboard**: The banner ad of dimensions 970×250 that loads on page load up and delivers a good impact for user experience is called a billboard.

3. **Interstitials**: The cross-section media adverts that assume control over the whole screen only seconds after webpage load up, launching from a banner advert of 1×1 pixel.

4. **Extending advert banners**: An advanced media advertisement which generates clicks or tap to reveal a large and interactive canvas.

5. **Pushdown**: An advert which pushes the content down rather than expanding over page content..

6. **Native infeed ads**: These ads are a new family of features from AdSense that helps publishers create and implement infeed native ads across their sites [13]. No CSS needed.

7. **The advantages of infeed promotions**: In-article promotions utilize top-notch promoter resources so they look extraordinary and give a superior client experience to your browser. They are advanced by Google to guarantee that they perform well on your article pages. On qualified sites, in-article promotions can likewise show content suggestions, making it simpler for your guests to discover a greater amount of your substance. Infeed promotions offer a superior client experience to your guests. They're a piece of the client's stream and match the look and feel of your site. They offer the chance to further adapt your pages by placing ads in new places i.e., inside your feeds. Finally, infeed promotions are perfect for versatility as they can assist you with better adapting to the smaller screen spaces available on cell phones [2].

8. **How infeed promotions differ from standard advertisements**: The article promotion techniques mix well with the experience of the browser. They show up inside your articles, utilize top-notch sponsor resources (for example Responsive Ads on the Display Network), and utilize a format that accommodates their situation and follows the reader's flow. Qualified distributers can likewise show content proposals, or "coordinated substance". Manual or programmed arrangement in addition to manually setting up these advertisements on your pages, you can now naturally put promotions on your pages with adequately long articles. "Auto InArticle" is another promotion position that shows up at common break points in your content and is enhanced to furnish your clients with a superior, and progressively local, advertisement experience. To begin, switch on this AdSense Lab to naturally embed an in-article local promotion inside your page.

9. **Manual or programmed arrangement**: Publishers can manually arrange infeed promotions or utilize the new auto infeed position. Turning on this AdSense Lab will naturally embed a pre-styled auto infeed local promotion inside your content-related feed. These promotions are shown by AdSense at ideal occasions to help increase income and give a decent client experience [14]. They just show up on very good-quality cell phones. The publishers likewise investigate the content and picture components of your feed to ensure they mix well with your substance. Local in-article promotions fit flawlessly in the middle of the passages of your pages for an improved reading experience.

10. **Display responsive ads**: Responsive promotions empower publicists to transfer separate imaginative resources, including features, portrayals, and pictures, that will at that point recombine to incorporate into the focal progression of the page or application [13,14]. The outcome is an ideal exhibition for the two publicists and distributers. Publicists have regularly observed a 15% expansion in reach at comparative execution contrasted with standard content promotions [14]. Likewise, promoters will have the option to tweak their creative adverts by transferring inventive resources that meet their image prerequisites. Responsive ads supplanted text ads on the Display Network beginning in 2017. Utilize this configuration for a simple method to make your promotions fit flawlessly inside the sites and applications individuals visit, while conveying a predictable encounter across devices. Provide your content, picture, and logo, and Google will structure your promotions to fit flawlessly across in excess of 2,000,000 applications and sites on the Google Display Network (GDN). Responsive promotions likewise open new local

stock so you can draw in buyers with advertisements that match the look and feel of the site or application they're utilizing [15].

11. **Promotion view**: Within the creation stream, publicists will have the option to see a review of delegate tests of their advertisements before they actuate them, empowering them to make changes before initiating their advertisement venture [14,15].

12. **Imaginative rendering**: The transferred resources will be rendered on the fly to incorporate into accessible stock. Responsive advertisements render in all GDN sizes and promotion designs – content, picture, and website. In the background, we take the sponsors' innovative resources and render them to best fit the distributer promotion spaces. Dynamic responsive promotions permit dynamic remarketing publicists to make, design, and run delightful unique advertisements on the Google Display Network, while still driving the most ideal exhibition. At the point when your responsive advertisement campaign is related with a feed (for example item feed), it will naturally be dynamic-enabled and can be utilized for dynamic remarketing purposes [16].

13. **Incredible performance**: Dynamic responsive advertisements easily provide the most ideal presentation. Publicists utilizing dynamic responsive advertisements see by and large 12% more transformations and a 17% increase in spending (at a minor 3% CPA increase) contrasted with their legacy dynamic remarketing promotions [16]. They're likewise future evidence – as new industry patterns develop around plans or configurations; Google consequently tries different things with adding these components to the advertisement format to convey an ideal presentation [14,16].

14. **Streamlined, yet incredible customization**: Create lovely advertisements effortlessly. For picture promotions, hues are naturally picked from your logo or can be chosen by the sponsors. Publishers will modify your advertisement from a chosen pool of best performing, outwardly engaging, and industry-explicit designs.

15. **Backdrop ads**: Backdrop ads are a lovely, responsive skin format for desktop and mobile web. The responsive structure looks incredible over a wide range of sites and screen sizes, working across work areas, tablets, and cell phones [17]. It incorporates a close button for a conscious client experience and is accessible automatically on the DoubleClick Ad Exchange.

16. **High-impact touch made automatically**: DoubleClick Bid Manager's backdrop format is a delightful, responsive touch design that looks extraordinary over a wide range of sites and screen sizes. It includes a close button for a respectful user experience and is accessible automatically on the DoubleClick Ad Exchange [17].

17. **When would it be a good idea for me to utilize it?** Use this configuration when you need to ensure individuals see your promotion. The digital equivalent of a billboard, Backdrop allows you to get your brand message, with plenty of room to make your point.

18. **Advertisement benefits**: Backdrop is simpler to use than most formats. Distributers are available to transact programmatically over Programmatic Guaranteed and other programmatic deal types.

19. **Dynamic image ads: Carousal**: Dynamic picture promotions give a lot of industry-explicit formats that feature the data clients need to reconnect with your image and convert. Advertisements are made in HTML5 and in ten distinctive promotion sizes, boosting reach across devices. Google currently suggests utilizing dynamic responsive advertisements, likewise recorded in the Ad Format Gallery, rather than this configuration since they offer the freshest, most adaptable, and lightweight imaginative answers for all powerful remarketing efforts [18].

20. **Promotion benefits**: Create excellent powerful picture advertisements adjusted to your image rules. The AdWords Ad Gallery empowers customization of various parts of the promotion, including background image/colors, call-to-action button, and others. These information sources are applied to all formats and to the ten distinctive advertisement sizes made consequently.

21. **Industry-explicit formats**: Dynamic remarketing is accessible to a wide range of organizations. It gives industry-explicit designs that feature the correct data for every client. For instance, while retail formats demonstrate a sponsor's item pictures from the Google Merchant Centre feed, there are additionally flight-explicit designs that show flight source and goal data [19].

22. **AdWords naturally improves your format**: AdWords applies your customization contributions to all designs, and runs tests out of sight to convey your best performing advertisements to clients. For promoters with exacting brand rules, they can choose explicit formats that will be a piece of the improvement pool, or select a solitary design to show to clients [19].

7.6 Advance the Display Network Campaigns

1. Comprehend the area of advertising.
2. GDN won't convey drives/deals at a similar effectiveness as your search campaigns.

In the domain of search, clients obviously recognize the issue they are attempting to fathom. When leveraging display, we can reach the same people, but not at the "moment of truth" – this bit of information drives GDN.

7.6.1 Identifying Campaign Objectives

The campaign will work at every phase of the advertising funnel by focusing on the needs of the channel and essential requirements of the client. Developers should pick the measurement metrics and set goals. Adverts need to be unique. And for this, these three things are required: *(a) top of funnel committed on CTR and location, (b) mid-funnel which will change the metrics, and (c) bottom of the funnel which calibrates the conversions* [20].

7.6.2 Manufacture the Ultimate Creativity

GDN focuses on apparatuses encourage the group of spectators, yet that is a long way from bringing deals to a close and legitimizing an immense raise toward the year's end [19]. The developer has to understand visitors with some reasonable objectives and keep them at the top of the priority list, comprehend the stage which needs to be focused, and realize your group of browsers all around to distinguish what will impact them. The developer has to sit with the promoter for setting up two sets of advertisements, one for testing and other as a back-up. Then if the advert runs well, then it is all set to go ahead.

7.6.3 Review the Checklist before Going to Upload

GDN checklists are as follows:

1. Check for campaign objectives setting.
2. Set the goal.
3. Check the target.
4. Check for unique creativity.
5. Make a good landing page.
6. Continue testing.

7.7 Remarketing Campaigns

Remarketing enables you to deposit a cookie via the browser of your website visitors. You will at that point have the option to show promotions focusing on those site guests at whatever point they visit different destinations additionally on GDN. Furthermore, it has been demonstrated to convey lower CPAs, so it's an incredible choice for sponsors working on a smaller spending plan [21].

Here's the manner by which to arrange an essential remarketing effort to target clients that have recently viewed the site:

1. Set remarketing and tags to all pages.

2. Create remarketing records from the crowd tab inside AdWords' shared library.

3. Within a maximum of two days, all visitors will populate which can be screened by the audiences tab in Adwords.

4. Once you have had 100 guests inside a period of 30 days, remarketing starts. To set up another campaign, click + *Campaign and select "Display Network only"*.

5. Select "**No marketing objective**" and "**All features**".

6. Click on a **target location, language, bid strategy, budget**, and **campaign name**.

7. Click **Save and continue**.

8. Name the first ad group and set default CPC bid. Click **Interests and remarketing** under *"Choose how to target your ads"*.

9. In the "Select a category" drop-down menu, choose **Remarketing lists**.

10. Choose one of the remarketing lists you previously created via the **audiences tab**. In this example, we will be targeting "all website visitors".

11. Click **Save and continue** if you want to create your ads, or **Skip ad creation** if you want to do this later.

12. Once your ads have been created, they're eligible to be served to all previous website visitors whenever they're browsing sites on the GDN.

7.7.1 Target Similar Traffic

To target similar traffic, set up another presentation campaign and select the "Like remarketing records" alternative inside the *Display Network* focusing on settings: Click the arrows » next to each similar audience to target. Similar audiences allow you to increase the reach of existing remarketing campaigns to drive new users to your site who have similar browsing habits as previous website visitors.

7.7.2 Focus on Proper Place for Adverts

Among all the targeting options accessible to promoters on the Google Display Network, the most ideal method for dealing with where your advertisements are being served is by utilizing overseen situations [22]. If you focus on techniques like interests or themes, Google is basically speculating about which destinations are applicable, which means there's a risk that your advertisements might be served on irrelevant sites. By focusing on positions, you're ready to choose precisely which destinations you need your promotions to be shown on, so it's a sheltered alternative for publicists simply fanning out into showcase publicizing. To set up placement targeting, navigate to the targeting settings of your display network campaign and choose *"placements"* from the *Ad targeting* drop down and then select specific placements from the list by searching for keywords or specific websites [23]. If you have a list of high-performing websites, one can begin to isolate them into individual advertisements. At that point you can all the more adequately deal with your offers on the situations that are giving the best return.

7.7.3 Look out for Mobile Games

Are you using keyword, topic, or interest targeting? Has it been a little while since you checked the situations your promotions are being served on? Provided that this is true, odds are that different mobile games are squandering a tremendous chunk of your financial limit. This is an issue that has shown up to a great extent because of children utilizing their parents' different mobile applications, which regularly leads to promotions being clicked on coincidentally [24]. How to check whether mobile games have been taking up a slice of your display budget: Navigate to showcase system campaign, click on the Display Network tab, and then Placements sub-tab.

7.7.4 Focus for Audience in the Market

In-market audiences are individuals whom Google has resolved to be most keen on what you bring to the table dependent on their browsing conduct and movement. This group of spectators involves clients who are effectively looking at and comparing products and services matching those you offer [25]. To check whether selective crowds are an effective targeting on technique for you, set up a campaign via the "Interests and remarketing" tab. Select In-market audiences from the drop down before checking which type of audience you'd like to target.

7.8 Check Gmail-Sponsored Advertisements

One of the most financially savvy kinds of showcase adverts is Gmail-sponsored advertisement (GSA). They show up inside the Gmail inbox, above messages. Intended to resemble a standard e-mail, once clicked they extend to a full promotion. In a typical case of how the advertisements show up in my inbox for Gmail, before being clicked, it will always come with *Ad* marked beside the mail in left side. Once clicked they will expand into the larger version of the product advertisement. Collapsed content-based text adverts and one extended advert are necessary. They can be made utilizing custom HTML. Sponsors are charged for every click. Note that the click is counted at whatever point somebody extends a promotion, not exactly when they navigate to the site from a promotion. Promoters will pay for the principal click. It's important to note that these promotions are just qualified to show up in standard Gmail accounts. They won't show up in accounts that are a part of a paid G Suite business account. For B2B publicists, it is worth remembering this, as you'll just have the option to arrive at business clients on their own Gmail accounts [26]. As far as focusing on alternatives, you can target by using any of these accompanying points:

1. Display keywords (this depends on the substance of the last 300 messages a client has gotten).
2. AdWords customer match.
3. Location focusing on (one for every search).
4. Topics (in view of the client's inbox).
5. Demographics (age/sexual orientation/parental status).

7.9 AdWords Conversion Setup

1. Go to *Tools* after logging into AdWords on Google.
2. Click on *Conversions*.
3. Click on the big red button marked Add Conversion and click on the choice of source to add (website, app, phone call, import).

7.9.1 Measurement of Conversion

Value: Here, we have choices for doling out our conversion a value. This is helpful for a couple of reasons. In the first place, it lets you attribute revenue and relative success to your AdWords campaigns. With the help of this information, one can compute things like when the visitor arrived at the advert and how much time he gave to it [27]. The return on advertisement spend (ROAS) and return on investment (ROI) can be measured. ROAS never consider any expenses other than the amount spent on adverts, but ROI calculates expenditures. Second, giving your conversion a value provides an opportunity to use a "target return on advertisement spend (ROAS)" portfolio (some time ago "adaptable") offering procedure, where the AdWords framework will fix offers to hit your ROAS objective. This is a helpful apparatus in the event that one has less time [28].

Here are the choices:

1. **Each time it occurs, the transformation activity has a similar worth**: With this decision, each time this conversion tag explores, it will be worth the same amount. This will supersede any exchange explicit qualities values passed by this tag. If you are not able to score leads, this alternative is useful for recognizing the general significance of two unique transformations [29]. For instance, somebody pursuing an e-mail rundown might be successful; however somebody making a record is more valuable than the business. So, we might assign an email list signup a value of $10 and an account creation a value of $20.

2. **The estimation of this change activity may shift**: Here, we're revealing to AdWords that this transformation may have various qualities depending on what item or administration a client chooses [30]. This alternative ought to be utilized for online business exchanges. We need to place in a default an incentive in the event that the tag doesn't explore appropriately; a great guideline is to utilize your normal your average order value. Utilizing this alternative will require customization with respect to your designer.

3. **Try not to assign a value**: This does precisely what you figure it does. *One never utilize this*, however if you don't especially think about the relative value of your conversions, don't hesitate to utilize it. Remember that you won't have the option to utilize that extravagant portfolio bidding strategy [31].

Count: Next, we have to decide what number of transformations to tally. We have two options:

1. **Every count**: Each click of transformation will be counted as a promotion [32]. This is the option to use if you are tracking sales, since you want to accurately track revenue. In this way, if somebody makes five buys after their click, you will see five transformations.

2. **One count**: This tally just a single transformation after an advertisement click. Pick this in case you're creating leads. If somebody unintentionally gives two structures, for instance, one may need to count one [33].

Windows of conversion: Here we mostly consider the advertisements that are clicked and viewed by impressions for a change to be qualified. Visitors who can be converted at the click are considered as a conversion, and if that conversion doesn't happen then you can't call it a conversion [34]. This depends on the business cycle starting from 1 day to 90 days. This is a very important setting where the window is set for 30 days and if a purchase or conversion happens on the 31st day then it won't be recorded. For this one can follow (*Tools >> Attribution >> Time Lag*).

Category: For selecting category, click on the suitable case from the drop-down box with a few alternatives like: (a) Other, (b) Buy/Sale, (c) Sign-Up, (d) Lead, (5) View of Key Page [26].

Attribution model: This is a moderately new yet incredible component for AdWords. We currently can decide how much credit every promotion click gets for a transformation. This choice is accessible for search and buying purposes at present [35]. It is a fantastic method of AdWords expression where generally we would be able to gauge achievement in terms of paid enquiry which drives changes [36]. For this there are some models to explain:

1. Last click: This is by default where the click on the last advertisement is given credit for 100% change.
2. First click: This is the opposite of last click. All credit goes to the primary adverts the client clicked.
3. Linear: The credit for conversion is equally spread out among all advertisements.
4. Time decay: More credit goes to that advertisement which is nearest to change. It uses a seven-day half-life to segregate the credit. So, adverts of eight days and those of one day will get the same credit if they have occurred before the change.
5. Position-based: This model gives 40% credit to the main click, 40% to the last click, and 20% to the rest of the clicks.
6. Data-driven: If you have enough transformation information, you can approach this unique sixth model. It utilizes your current information to choose how to scatter credit among your campaigns. To be qualified, Google says you need "in any event 20,000 ticks and a change activity must have at any rate 800 transformations inside 30 days". If your numbers plunge underneath a specific limit, you can never again utilize the model.

Which model you pick will depend significantly on your association and your objectives. Last click is the most moderate, so on the off chance that you don't have much time to deal with your account, then the above-mentioned attribution techniques can be worthy alternatives [37]. Then again, attribution modeling overlooks different campaigns that additionally earned that transformation. As a rule, people don't simply tap on one advertisement and buy. There are regularly numerous touchpoints on the way to choosing. In the event that you need that information, a direct model or time-rot model are decent compromises. In the event that you need to organize the underlying connection or give more credit to keywords that fall higher in the conversion path (and are commonly increasingly conventional, costly terms), a first-click or position-based model may suit your needs, as somebody who works in AdWords consistently may have the advantage of more opportunities to dissect and follow up on this information. Something else to remember is click-assisted conversions, which measure the quantity of clicks that aided a transformation way. In the event that you utilize a model that considers these snaps, similar to a position-based model, you don't have to utilize this measurement for estimation. Remember, this only works inside the AdWords framework; it doesn't consider different channels. For that, you should utilize Google Analytics to settle on a bigger choice about where paid inquiry fits into your computerized promoting blend [38].

Implementation: Eventually, whether you track on a page load or a click will boil down to how your site is managed. In the event that you have a designer, you can send them the tag to actualize by clicking "Email guidelines and tag", which will give them the tag and directions for execution. The label needs to go between your *page's <body> and </body> labels, or you can include it utilizing Google Tag Manager.* At the point when it's up, you are headed to estimating accomplishment in your AdWords account!

7.10 Conclusion

Utilizing multivariate testing as a strategy for website streamlining is an incredible technique for gathering visitor and user data that gives detailed insights for understanding complex client conduct. The information revealed in multivariate testing expels uncertainty and vulnerability from site advancement. Continuously testing, implementing winning variations and building off of testing insights can lead to significant conversion gains. Remarketing is a shrewd speculation that could build your primary concern – in the event that you do it right. Nothing will tank your remarketing system quicker than terrible administration. While it appears to be moderately basic, the best outcomes originate from the highest audience segmentation and best bidding strategies. Google AdWords could be used to help internet business exercises. The utilization of Google AdWords is helpful to expand deals. It has numerous advantages that could be used to build the quantities of visits to the sites, the quantities of new clients, and to extend the business. Because of the advantages and focal points, individuals are recommended to utilize Google AdWords to support their activities and to increase the sales.

References

1. Johansson, C., & Wengberg, P. (2017). *Dynamic retargeting: The Holy Grail of marketing?* (Dissertation), p. 48. Retrieved from http://urn.kb.se/resolve?urn=urn:n bn:se:uu:diva-325621.

2. Tricahyadinata, I., & Za, S.Z. (2017). An analysis on the use of google adwords to increase E-commerce sales. *SZ Za and I. Tricahyadinata (2017) International Journal of Social Sciences and Management, 4*, 60–67.

3. Anagnostopoulos, A., Broder, A.Z., Gabrilovich, E., Josifovski, V., & Riedel, L. (2007, November). Just-in-time contextual advertising. In *Proceedings of the sixteenth ACM conference on information and knowledge management* (pp. 331–340).

4. Dias, M.B., Locher, D., Li, M., El-Deredy, W., & Lisboa, P.J. (2008, October). The value of personalised recommender systems to e-business: A case study. In *Proceedings of the 2008 ACM conference on recommender systems* (pp. 291–294).

5. Mehta, A., Saberi, A., Vazirani, U., & Vazirani, V. (2007). AdWords and generalized online matching. *Journal of the ACM (JACM), 54*(5), 22-es.

6. Das, S., Nayyar, A., & Singh, I. (2019). An assessment of forerunners for customer loyalty in the selected financial sector by SEM approach toward their effect on business. *Data Technologies and Applications, 53*(4), 546–561.

7. Subhankar, D., & Anand, N. (2019, May). Digital sustainability in social media innovation: A microscopic analysis of Instagram advertising & its demographic reflection for buying activity with R. In *1st International scientific conference "Modern Management Trends and the Digital Economy: From Regional Development to Global Economic Growth" (MTDE 2019)*. Atlantis Press.

8. Singh, I., Nayyar, A., Le, D.H., & Das, S. (2019). A conceptual analysis of internet banking users' perceptions. An Indian perceptive. *Revista ESPACIOS, 40*(14), 1–17.

9. Berkowitz, D., Allaway, A., & D'souza, G.I.L.E.S. (2001). The impact of differential lag effects on the allocation of advertising budgets across media. *Journal of Advertising Research, 41*(2), 27–27.

10. Bettman, J.R. (1979). *Information processing theory of consumer choice*. Addison-Wesley Pub. Co., Boston, MA.

11. Billsus, D., & Pazzani, M.J. (1998, July). Learning collaborative information filters. *ICML, 98*, 46–54.

12. Bleier, A., & Eisenbeiss, M. (2015). The importance of trust for personalized online advertising. *Journal of Retailing, 91*(3), 390–409.

13. Franco, C.E., & Regi, B.S. (2016). Advantages and challenges of e-commerce customers and businesses: In Indian perspective. *International Journal of Research–Granthaalayah, 4*(7), 7–13.

14. Das, S; Mondal, S.R.; Sahoo, K.K; Nayyar, A. & Musunuru, K. (2018). Study on impact of socioeconomic make up of Facebook users on purchasing behavior. *Revista Espacios, 39*, 28–42. Retrieved from http://www.revistaespacios.com/a18v39n33/18393328.html

15. Mondal, S., Das, S., Musunuru, K., & Dash, M. (2017). Study on the factors affecting customer purchase activity in retail stores by confirmatory factor analysis. *Revista Espacios, 38*, 30–55. Retrieved from http://www.revistaespacios.com/a17v38n61/17386130.html

16. Mondal, S., Mall, M., Mishra, U.S., & Sahoo, K. (2017). Investigating the factors affecting customer purchase activity in retail stores. *Revista ESPACIOS, 38*(57), 22–44.

17. Kumar Sahoo, K., & Mondal, S. (2016). An analysis of impact of electronic customer relationship management (e-CRM) on service quality of E-retail stores: A study of Bhubaneswar. *Research Revolution, 2*, 10–12.

18. Mondal, S., & Sahoo, K.K. (2020). A study of green building prospects on sustainable management decision making. In *Green building management and smart automation* (pp. 220–234). IGI Global.

19. Das, S., & Nayyar, A. (2020). Effect of consumer green behavior perspective on green unwavering across various retail configurations. In *Green marketing as a positive driver toward business sustainability* (pp. 96–124). IGI Global.

20. Gupta, D.K., Jena, D., Samantaray, A.K., & Das, S. (2019). HRD climate in selected public sector banks in India. *Revista ESPACIOS, 40*(11), 14–20.

21. Singh, S., & Das, S. (2018). Impact of post-merger and acquisition activities on the financial performance of banks: A study of Indian private sector and public sector banks. *Revista Espacios Magazine, 39*(26), 25.

22. Jain, S., Jain V., & Das, S. (2018). Relationship analysis between emotional intelligence and service quality with special evidences from Indian banking sector. *Revista ESPACIOS, 39*(33).

23. Singh, S., Mondal, S., Singh, L.B., Sahoo, K.K., & Das, S. (2020). An empirical evidence study of consumer perception and socioeconomic profiles for digital stores in Vietnam. *Sustainability, 12*(5), 1716.

24. Singh, L.B., Mondal, S.R., & Das, S. (2020). Human resource practices & their observed significance for Indian SMEs. *Revista ESPACIOS, 41*(7). Retrieved from http://www.revistaespacios.com/a20v41n07/20410715.html

25. Sharma, E., & Das, S. (2020). Measuring impact of Indian ports on environment and effectiveness of remedial measures towards environmental pollution. *International Journal of Environment and Waste Management, 25*(3), 356–380. doi:10.1504/IJEWM.2019.10021787

26. Berkowitz, D., Allaway, A., & D'Souza, G. (2001). Estimating differential lag effects for multiple media across multiple stores. *Journal of Advertising, 30*(4), 59–65.

27. McElfresh, C., Mineiro, P., & Radford, M. (2008). *U.S. Patent No. 7,373,599.* Washington: U.S. Patent and Trademark Office.

28. Gupta, A. (2014). E-commerce: Role of E-commerce in today's business. *International Journal of Computing and Corporate Research, 4*(1), 1–8.

29. Johnston, M.P. (2017). Secondary data analysis: A method of which the time has come. *Qualitative and Quantitative Methods in Libraries, 3*(3), 619–626.

30. Das, S. (2020). Innovations in digital banking service brand equity and millennial consumerism. In *Digital Transformation and Innovative Services for Business and Learning* (pp. 62–79). IGI Global.

31. Mondal, S.R. (2020). A systematic study for digital innovation in management education: An integrated approach towards problem-based learning in Vietnam. In *Digital innovations for customer engagement, management, and organizational improvement* (pp. 104–120). IGI Global.

32. Nadanyiova, M., & Das, S. (2020). Millennials as a target segment of socially responsible communication within the business strategy. *Littera Scripta, 13*(1), 119–134. doi: 10.36708/Littera_Scripta2020/1/8

33. Mehta, A., Saberi, A., Vazirani, U., & Vazirani, V. (2007). AdWords and generalized online matching. *Journal of the ACM (JACM), 54*(5), 22-es.

34. Niranjanamurthy, M., Kavyashree, N., Jagannath, S., & Chahar, D. (2013). Analysis of e-commerce and m-commerce: Advantages, limitations and security issues. *International Journal of Advanced Research in Computer and Communication Engineering, 2*(6), 2360–2370.

35. Rajaraman, V. (2001). 6. Electronic data interchange and XML. *RESONANCE, 5*(10), 13–23.

36. Riggins, F.J., & Rhee, H.S. (1998). Toward a unified view of electronic commerce. *Communications of the ACM, 41*(10), 88–95.

37. Treese, G.W., & Stewart, L.C. (2003). *Designing systems for internet commerce.* Boston, MA: Pearson Education, Inc.

38. Vladimir, Z. (1996). Electronic commerce: Structures and issues. *International journal of electronic commerce, 1*(1), 3–23.

Index

A

Accelerated mobile pages (AMP), 124
Ad Promotion preview, 165
Advertisement by videos, 11
AdWords, 141; *see also* Multivariate
      testing remarketing and
      AdWords
  broad match type, 141
  conversion setup, 195–198
  exact match type, 142
  keyword planner, 65
Ahrefs toolset, 76
AI, *see* Artificial intelligence
AIDA, *see* Attention, interest, desire,
      and action; Attention, interest,
      desire, and action
Alt tag, 60–61
Alt text, 41
AMP, *see* Accelerated mobile pages
The Anatomy of a Large-Scale
      Hypertextual Web-Search
      Engine, 13
Anchoring, 146
Anchor text, 77, 92
AnswerThePublic, 64
Antiquated search, 65
Apache, 56
Archie, 12
Article
  accommodation, 98, 99
  destinations, 98
  directory submission, 16
  expansion, 99
Artificial intelligence (AI), 184
Attention, interest, desire, and action
      (AIDA), 11, 187
Attribution modeling
  customized algorithmic
      attribution, 23
  first-touch attribution, 18, 19
  full-path (Z-Shaped) attribution,
      22, 24
  last non-direct touch attribution, 19–20

  last-touch (opportunity creation
      touch) attribution, 18–20
  last (insert marketing channel) touch
      attribution, 21
  lead-generation touch attribution,
      18, 19
  linear attribution, 21–22
  position attribution, 22, 23
  position extended attribution,
      22, 23
  time decay attribution, 22
Autocomplete tools, 63–64
Auto InArticle, 189

B

B2B, *see* Business to business
B2C, *see* Business to customer
      remarketing
Backdrop ads, 190
Backlinks, 16, 91–92
  dofollow *vs.* nofollow, 93–94
  getting backlinks, 93
  importance of, 92
Bids, 142
Billboard, 188
Bing Ads, 95, 141
Bing Webmaster Tool, 65
Blog directories, 96
Blogging, 147
Business listings, 110
Business to business (B2B)
  awareness, 130, 131
  structure, 18
Business to customer (B2C)
      remarketing, 184

C

Call to action (CTA), 40, 163, 164
Carousal, 191
CCMS, *see* Component content
      management system
Chain mode, 127

Printed in the United States
by Baker & Taylor Publisher Services